T0140113

Metal Matrix Composites

Antonio Contreras Cuevas
Egberto Bedolla Becerril
Melchor Salazar Martínez
José Lemus Ruiz

Metal Matrix Composites

Wetting and Infiltration

 Springer

Antonio Contreras Cuevas
Instituto Mexicano del Petróleo
Ciudad de México, México

Egberto Bedolla Becerril
Universidad Michoacana de San
Nicolás de Hidalgo
Instituto de Investigación
en Metalurgia y Materiales
Morelia, Michoacán, México

Melchor Salazar Martínez
Clúster Politécnico Veracruz -IPN
Papantla de Olarte, Veracruz, México

José Lemus Ruiz
Universidad Michoacana de San
Nicolás de Hidalgo
Instituto de Investigación
en Metalurgia y Materiales
Morelia, Michoacán, México

ISBN 978-3-030-06312-2   ISBN 978-3-319-91854-9   (eBook)
https://doi.org/10.1007/978-3-319-91854-9

This Springer imprint is published by the registered company Springer Nature Switzerland AG
The registered company address is: Gewerbestrasse 11, 6330 Cham, Switzerland

Preface

Metal matrix composites (MMC) have been used in industrial forms for several decades; however, they continue improving. MMC usually consist of two different phases: one phase is a matrix (usually a metal) and the other is a reinforcement (usually a ceramic or a metal). These two phases are mixed by an appropriate technique in order to obtain a homogeneous material with specific properties that are different from the two monolithic materials.

MMC are mainly used in the automotive, sporting, and spatial industries, although some are used in various high-temperature applications. MMC may be designed and produced to obtain certain mechanical, electrical, and thermal properties for specific applications. For example, high-temperature MMC can be used in automotive engine components, turbines, and spaceships, among others.

Among ceramic, polymeric, and metallic matrices, the most widely used and studied has been MMC. Only a few years ago, MMC was studied by only a handful of researchers worldwide; now, MMC has become one of the most popular research subjects in materials science. Particular attention has been focused in Al and Mg alloys used as matrices, which are widely used in metallic matrix composites (MMC). The advantages of magnesium and its alloys used as a composite's matrix are their high specific strength and stiffness, good damping capacities, and dimensional stability.

There are many books about composite materials, covering all aspects related to processing, characterization, and applications of MMC. However, this book, in addition to covering all these aspects, includes a wide review of experimental results in wetting, fabrication techniques, thermodynamics, kinetics, corrosion, wear, and welding in relation to MMC. All this research was conducted by our group of researchers over more than 30 years at Instituto de Investigación en Metalurgia y Materiales from Universidad Michoacana de San Nicolás de Hidalgo (UMSNH) in the city of Morelia, México. The great quantity of experimental results obtained by this research group related to MMC, including several systems of composites, provides the motivation to write this book. This book provides the basic theory of MMC and results in wetting and processing of several systems of composites, such as Al-alloys/TiC, Mg-alloys/TiC, Cu/TiC, Al/AlN, Al-alloys/SiC, Mg/AlN, Ni/TiC, Mg/SiC, and Ni/Al$_2$O$_3$, among others. It is intended to be useful for students working in MMC, because it covers all the fundamental aspects in wetting, interfacial reactions, processing, mechanical characterization, welding, corrosion, and wear of MMC; it should also prove useful for researchers because it includes a complete review of experimental results in all the topics mentioned above.

This book contains seven chapters. The first chapter gives a general introduction of MMC. The second chapter contains all the main theories about wetting, one of the most important phenomena to be considered in the fabrication of composites. Moreover, it includes the main experimental results of wetting and includes a complete study on the interfacial characterization between the matrix and reinforcement of the fabricated composites. The third chapter describes all the main techniques existing for fabrication of composites and focuses overall on infiltration techniques. The fourth chapter contains the main results and characterization of different composite systems, including microstructural, mechanical, thermal, and electrical, that the group has worked on for more than 25 years. The fifth chapter presents all the fundamentals of joining composites, the main experiment carried out and its methods, including welding and brazing, and the results obtained by the group utilizing different composite systems. The sixth chapter contains results of the corrosion behavior of some composites exposed to aqueous media. The main research performed in corrosion of composites uses TiC and SiC as reinforcements and Al, Ni, and some Al-Cu$_x$, Al-Mg$_x$, and Al-Cu-Li alloys as matrices. The corrosion behavior of MMC is of great importance when determining how and when to use it in particularly corrosive environments. The last chapter presents the fundamental aspects of the wear of composites and some results related to wear of the TiC/Mg-AZ91E system.

Ciudad de México, México   Antonio Contreras Cuevas
Morelia, Michoacán México   Egberto Bedolla Becerril
Papantla de Olarte, Veracruz México   Melchor Salazar Martínez
Morelia, Michoacán México   José Lemus Ruiz
June 2018

Credits to Publishers and Collaborators

Credits to Publishers

We would like to acknowledge the following publishers who granted the permission to reproduce the figures and photographs used in this book:

- Cambridge University Press
- Elsevier
- Elsevier Science Publishers
- John Wiley and Sons
- Scitec Publications
- Springer Nature
- Springer Publishing Company
- Taylor & Francis
- Trans Tech Publications

Credits to Our Work Group

I would like to give special thanks to the following students (now researchers) who contributed to the experimental results via their theses and articles. All of them were students in the Instituto de Investigación en Metalurgia y Materiales from Universidad Michoacana de San Nicolás de Hidalgo, México, and they were part of our group working in composite materials managed by Dr. Egberto Bedolla for more than three decades.

Dra. Ena A. Aguilar Reyes
Dra. Yadira Arroyo Rojas Dasilva
Dra. Alejandra Reyes Andres
Dr. Carlos A. León Patiño

Dr. Apolinar Albiter Hernández
Dr. Víctor H. López Morelos
Dr. Ricardo Morales Estrella
Dr. Lázaro Abdiel Falcón Franco
M.C. Carlos Arreola Fernández
M.C. Omar R. Zalapa Lúa
Dr. Rafael García Hernández
Dr. Leonel Ceja Cárdenas
Dra. Ana Lizeth Salas Villaseñor
Dra. Josefina García Guerra
M.C. Raúl Alejandro Pulido Aguilar
M.C. Estefania Ortega Silva
M.C. Gustavo Castro Sánchez
M.C. Rodrigo Alan Martínez Molina

Credits to External Institutions

The authors also would like to thank the Department of Mining and Metallurgical Engineering in McGill University, Montreal, Canada, led by Prof. Robin A. L. Drew, for all the support in the lab to carry out the wetting tests and some students of our group who performed research for their doctoral thesis.

Acknowledgments

Dr. Antonio Contreras Cuevas would like to thank Dr. E. Bedolla Becerril, Dr. J. Lemus Ruiz, and Dr. M. Salazar Martínez who accepted the challenge to write this book containing all experimental results of our research group and collaborating as coauthors of this book. Dr. E. Bedolla Becerril, as leader of the group, also wishes to acknowledge his past and present graduate students, research associates, and his colleagues who have worked on the different topics of this book, whose interesting results are used extensively in this book. The authors would like to thank the Instituto de Investigación en Metalurgia y Materiales from Universidad Michoacana de San Nicolás de Hidalgo (UMSNH) in Morelia México and the Department of Mining and Metallurgical Engineering from McGill University in Canada for its support in the experimental work. The authors also acknowledge the guidance and support of Professor R.A.L. Drew during all wetting experiments. Finally, the authors are grateful for the financial support received from CONACYT (Consejo Nacional de Ciencia y Tecnología).

Contents

Abbreviations

AC	Alternating current
AE	Auxiliary electrode
AFM	Atomic force microscopy
Al	Aluminum
Al_2O_3	Aluminum oxide
Al_4C_3	Aluminum carbide
AlN	Aluminum nitride
Ar	Argon
ASME	American Society of Mechanical Engineering
ASTM	American Society for Testing and Materials
AZ91E	Magnesium alloy
BF-STEM	Bright-field scanning transmission electron microscopy
CMC	Ceramic matrix composites
CR	Corrosion rate
CTE	Coefficient of thermal expansion
Cu	Copper
CVD	Chemical vapor deposition
CVI	Chemical vapor infiltration
CVN	Charpy V-notch
DC	Direct current
DEA	Direct electric arc
EB	Electron beam
E_{corr}	Corrosion potential
EDS	Energy-dispersive spectroscopy
EDX	Energy-dispersive X-ray
EIS	Electrochemical impedance spectroscopy
EPMA	Electron probe microanalysis
FFT	Fast Fourier transform
GMAW	Gas metal arc welding
HAADF	High-angle annular dark-field

HAZ	Heat-affected zone
HIP	Hot isostatic pressing
HRC	Hardness Rockwell C
HRTEM	High-resolution transmission electron microscopy
HVOF	High-velocity oxy-fuel
IEA	Indirect electric arc
IIW	International Institute of Welding
IRMS	Current root mean square
LI	Localization index
LM	Liquid metallurgy
LPPD	Low-pressure plasma deposition
LPR	Linear polarization resistance
MA	Mechanical alloying
Mg	Magnesium
Mg_2Si	Magnesium silicide
$MgAl_2O_4$	Aluminum magnesium spinel
MgO	Magnesium oxide
MGW	Modified gas welding technique
MIG	Metal inert gas
MMC	Metal matrix composite
MML	Mechanical mixed layer
Mn	Manganese
MOR	Modulus of rupture
Nb	Niobium
Ni	Nickel
OCP	Open-circuit potential
PACVD	Plasma-assisted chemical vapor deposition
PAS	Plasma-activated sintering
PC	Polarization curves
PECS	Pulse electric current sintering
PI	Pressureless infiltration
PIC	Pressure infiltration casting
PM	Powder metallurgy
PTLPB	Partial transient liquid-phase bonding
PVD	Physical vapor deposition
RE	Reference electrode
RS	Rapid solidification
SAED	Selected area electron diffraction
SC	Stir casting
SCE	Saturated calomel electrode
SEM	Scanning electron microscopy
SHS	Self-propagating high temperature
Si	Silicon
Si_3N_4	Silicon nitride

SiC	Silicon carbide
SiO_2	Silicon dioxide
SPS	Spark plasma sintering
TC	Thermal conductivity
TEM	Transmission electron microscopy
TGA	Thermogravimetric analyzer
Ti	Titanium
TiC	Titanium carbide
TiO_2	Titanium oxide
TLP	Transient liquid phase
UTS	Ultimate tensile strength
WC	Tungsten carbide
WDS	Wavelength-dispersive spectroscopy
WE	Working electrode
XD	Exothermic dispersion
XRD	X-ray diffraction
YS	Yield strength
Y-TZP	Yttria partially stabilized zirconia

Symbols

%	Percentage
°C	Celsius degrees
μm	Micron
A	Apparent area
A_r	Real area
At.%	Atomic percent
cm	Centimeter
CTE	Coefficient of thermal expansion
E	Elastic modulus
E_a	Activation energy
EN	Electrochemical noise measurements
ER	Electrical resistivity
F	Force
g	Grams
GPa	Gigapascal
h	Hour
Hv	Hardness Vickers
I_{corr}	Current density
I_{lim}	Limiting current density
J	Joules
K	Kelvin degrees
KHz	Kilohertz
KJ	Kilojoules
K_{lc}	Fracture toughness
KN	Kilonewton
Kw	Kilowatts
l	Liquid
m	Meter
M	Molar
min	Minute

mJ	Millijoule
mm	Millimeter
MPa	Megapascal
Q	Worn volume per distance
r	Roughness factor
R	Universal gas constant
R_n	Noise resistance
R_p	Polarization resistance
rpm	Revolution per minute
S	Aspect ratio
t	Time
T_{inf}	Infiltration temperature
T_m	Melting temperature
T_s	Sintering temperature
V	Worn volume
V_m	Volumetric fraction of the matrix
Vol	Volume
Vol.%	Volume percent
V_r	Volumetric fraction of reinforcement
W	Applied standard load
W_a	Adhesion work
W_c	Cohesion work
Wt	Weight
Wt.%	Weight percent
X	Wetting perimeter
α	Alpha phase
ΔG	Gibbs free energy
θ	Contact angle
θ_a	Advancing contact angle
θ_{hyst}	Hysteresis contact angle
θ_r	Receding contact angle
σ_i	Current noise standard deviation
σ_v	Potential noise standard deviation
Υ_{LV}	Surface energy of liquid-vapor
Υ_{SL}	Surface energy of solid-liquid
Υ_{SV}	Surface energy of solid-vapor
Φ	Apparent contact angle

Chapter 1
Introduction

1.1 Composite Materials

Composite materials in the last decades have gained considerable importance due to their high application in the automotive, aeronautics, aerospace, and electronics industries. However, these materials have been used by human culture since ancient times, such as "adobe" (old material used to build houses), which is a material made of mud and straw, as well as concrete made of cement and gravel; at present, this material is widely used in the construction industry. In nature, there are also composite materials, without the need to manufacture them, such as the bone constituted by hard inorganic crystals called hydroxyapatite and collagen as the matrix; wood is another natural composite formed by fibrous chains of cellulose enclosed by a matrix of lignin.

The composites are obtained by joining two or more materials that differ in shape and chemical composition and are insoluble with each other. The composites acquire properties that are not possible to obtain in the precursor materials. These are formed by a continuous phase called matrix, which may be metallic, ceramic, or polymeric, as well as a reinforcing phase dispersed in the matrix, which may be in the form of short fibers, long fibers, or particles [1]. Generally, the components are different in properties, being one light, strong, hard and fragile, and the other can be tough and ductile. Figure 1.1 schematizes a composite material composed by the reinforcement, matrix, and interface (metal-ceramic).

MMC can be classified according to reinforcement, and the matrix used is shown in Fig. 1.2. Currently, one of the most used composite materials are composites made of polymer matrix (PMC), since in the manufacture of these, it used lower temperatures than in the case of materials of metallic or ceramic matrix, the reason why it is made cheaper; however, the MMC can be used where higher temperatures are required as it has higher thermal stability.

© Springer Nature Switzerland AG 2018
A. Contreras Cuevas et al., *Metal Matrix Composites*,
https://doi.org/10.1007/978-3-319-91854-9_1

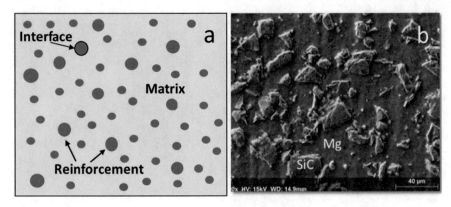

Fig. 1.1 (**a**) Scheme of a composite material and (**b**) SEM image of Mg/SiC composite

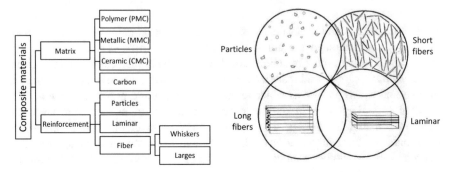

Fig. 1.2 Classification of the composite materials according to the type of matrix and reinforcement

1.2 Metal Matrix Composites (MMCs)

MMCs have attracted attention in recent years due to their good physical and mechanical properties for application in the aerospace industry (in turbine components, in mechanical subsystems, and in the manufacture of important sections of the fuselage) at present, and due to its low density and good mechanical properties, the automotive industry has increased the use of these materials [1]. In the same way, composites with high reinforcement contents are being studied and used in the electronics industry since they offer good mechanical properties, low coefficient of thermal expansion, and high thermal conductivity. Although metal matrix composite materials offer higher specific strength and stiffness, as well as higher operating temperatures than monolithically materials, the cost of production is still high, and it only used where the most important is the saving of the weight as in the aerospace and military industry. The land transportation industry is one of the sectors that most uses the MMCs including the cars and trains, although the automotive industry is the one that most uses this type of materials; perhaps the best examples of manufactured parts are components for brake discs, cylinders, pistons, shaft drive, among others.

The MMCs consist of a continuous phase called matrix, which must be metallic, which surrounds the discontinuous phase called reinforcement, and which may be in the form of continuous fibers, whiskers, or particles. Generally, the reinforcement phase is hard and brittle with high modulus of elasticity, and the matrix is usually ductile and tenacious; the matrix is responsible for transmitting the load to the interface and reinforcement, keeping the type of reinforcement spaced and oriented according to needs; and in the same way protects the reinforcement of the environment; in this way a homogeneous structure is obtained, with good values of stiffness, resistance, density, resistance to corrosion, hardness, good wear resistance mainly at low loads, low coefficients of thermal expansion, and better creep resistance at high temperatures than conventional alloys, and in some cases can obtain good thermal and electrical conductivity, depending on the matrix and the reinforcement, for example, aluminum and silicon carbide which have a good thermal conductivity [2, 3]. Long fibers provide the composite with better strength and modulus of elasticity properties than those made from short fibers and particles, but the long fibers have higher costs than the short fibers, and therefore the use of a fiber type or particle depends on the engineering needs.

Long fibers Composites with long fibers offer good mechanical properties such as modulus of elasticity and ultimate tensile strength, especially when load applied is parallel with the direction of fiber; this material is considered anisotropic and its properties depend on the direction of the material; in case of materials reinforced with particles, although their mechanical properties are smaller than those reinforced with long fibers, its properties are equal in any direction due to their isotropic characteristics.

Mechanical properties, such as tensile strength, hardness, modulus of elasticity, and wear resistance, increase with increasing volume percent of reinforcement, while ductility, toughness, and thermal conductivity are diminished with increasing reinforcement.

1.3 Types of Matrices and Reinforcement

Particle reinforcements are more commonly used in metal matrix composites because they are more economical to manufacture and greater isotropy is obtained in the properties of the finished product. At present the most used reinforcements in the form of particles are silicon carbide (SiC); titanium carbide (TiC); boron carbide (B$_4$C); titanium diboride (TiB$_2$); oxides such as alumina (Al$_2$O$_3$), titanium oxide (TiO$_2$), silicon oxide (SiO$_2$), or zirconium oxide (ZrO$_2$); and aluminum nitride (AlN) because its good thermal conductivity is being studied for use in electronic devices [4]; at present the use of intermetallic particles is progressively increased, as well as metallic reinforcements such as tungsten (W) and molybdenum (Mo) among other reinforcements. The percentage by volume of reinforcement can vary from very low percentages to 50%; however, for the electronics industry, compounds with amounts greater than 50% can be made and still are considered MMC since the continuous phase encloses reinforcement particles.

The matrix function as already said is to keep the reinforcements bonded, thanks to their cohesive and adhesive characteristics, as well as to transmit the load to the interface and the reinforcement; consequently, it is of fundamental importance to obtain a good adhesion between the matrix and the reinforcement.

Generally, it is thought that metal matrix composites should be fabricated with light metal matrices; however, the selection of this matrix depends on the use that will be given to the MMC. At present the metallic matrixes used are aluminum, iron, titanium, nickel, and magnesium, among others, underlining among all that of aluminum and its alloys, since they have low weight, good resistance to corrosion, and low melting point. Although the density of magnesium is approximately 2/3 of the aluminum, magnesium is less used in the manufacture of MMC, so it is very important to carry out research of this material since one of the reasons for the use of MMC is to manufacture materials with low densities but with mechanical properties equal or better than the conventional materials, and since, today, the main challenge for humanity is the preservation of the environment which starts with the reduction of fuel consumption by reducing the weight of transportation vehicles [5].

Different routes are used to produce MMC, such as the casting method with agitation, which is the most economical and easily applicable to high production volumes; however, it has some drawbacks such as nonhomogeneous distribution of reinforcements, porosity, and relatively low reinforcement volumes. Other technologies are powder metalurgy, infiltration and in situ manufacture at elevated temperatures. Infiltration methods are widely used, and could be infiltration with and without pressure. Infiltration with pressure involves a mechanical pressure used to force the liquid metal to infiltrates the preform; this method is known as squeeze casting, and also gas pressure infiltration technique is used. The MMC may be fabricated by pressureless infiltration route using the capillary phenomenon, that is to say, the infiltration of the liquid metal in the porous preforms infiltrates without having to apply any pressure; however, a good wettability between the matrix and the reinforcement used is required. But in those cases where there is no good wettability, some techniques like addition of alloying elements in the matrix and coatings applied to reinforcement are used. The reinforcement is coated with another material, so that the wetting occurs, and in many cases the coating serves as a barrier between the reinforcement and the matrix, so that undesirable reactions do not occur among matrix and reinforcement which weaken the composite, such as aluminum reinforced with fibers or particles with carbon contents such as SiC or TiC and graphite that give rise to products such as aluminum carbide, which easily degrades in humid environments. An advantage of the method (infiltration of porous preforms) is that pieces generally has dimensions and shapes close to those required, and the disadvantage is that the method is expensive since it generally requires to sinter preform prior to infiltration, although investigations using the preforms in green report good results on its mechanical properties [4].

This book is intended for all those students or researchers who are interested in the manufacture of metallic matrix composites in which the research experience of this working group is transferred. Chapter 2 reviews all the parameters related to wettability of liquid metals in ceramic substrates, since the wettability is fundamental in the manufacture of MMC. Chapter 3 gives a brief description of the fabrication processes of MMC, but it will emphasize mainly in the infiltration processes, including a short description of the governing physical phenomenon. Chapter 4 contains the main results about characterization of different composite systems in which the group had worked for more than 25 years including a microstructural, mechanical, thermal, and electrical characterization mainly. In Chap. 5 the underlying science of joining the MMC is described. Some specific examples of production of joining MMC produced by the authors are given to illustrate the key factors involved. The properties of the MMC/metal ensembles can be related directly to the joining technique as well as the parameter and its effect in thermodynamic and mechanism involved, which must be controlled during joining processes. Chapter 6 contains results about corrosion behavior of some composites exposed to aqueous media. The main research was performed in corrosion of composites using TiC as reinforcement and Ni, Al-Mg, Al-Cu, and Al-Cu-Li alloys as matrixes. Chapter 7 contains the reviewing research work on the wear of composite materials considering the different type of tests currently used, the mechanisms of wear materials, and a comparison of results obtained when different techniques used the same material.

References

1. Mordike BL, Lukáč P (2001) Interfaces in magnesium-based composites. Surf Interface Anal 31 (7):682–691
2. Li S et al (2014) Thermophysical properties of SiC/Al composites with three dimensional interpenetrating network structure. Ceram Int 40(5):7539–7544
3. Mizuuchi K, Inoue K, Agari Y et al (2012) Processing of Al/SiC composites in continuous solid–liquid co-existent state by SPS and their thermal properties. Compos Part B 43:2012–2019
4. Bedolla E, Ayala A, Lemus J, Contreras A (2015) Synthesis and thermo-mechanical characterization of Mg-AZ91E/AlN composites. In: 10th international conference on magnesium alloys and their applications, Jeju, Korea
5. Nguyen Q, Sim Y, Gupta M et al (2014) Tribology characteristics of magnesium alloy AZ31B and its composites. Tribol Int 82:464–471

Chapter 2
Wettability

2.1 Wettability of Ceramics by Metals

For the processing of MMC's composites, the liquid metal must wet the ceramic phase. In the manufacture of composite materials by infiltration techniques, the liquid matrix, metal, or alloy must be in contact with the reinforcing material (fibers or particles) for infiltration followed by solidification. The degree of infiltration of the liquid metal into the preform will depend on the degree of wetting of the metal on the reinforcing material. The degree of wettability is given by the contact angle (θ) which is formed by a liquid drop resting on a solid substrate as is shown in Fig. 2.1.

The relations of the vectors of the surface forces are given by the equation:

$$\gamma_{SV} - \gamma_{SL} = \gamma_{LV} \, \text{Cos} \, \theta \qquad (2.1)$$

where γ_{SV}, γ_{SL}, and γ_{LV} are the surface energies of the solid-vapor, solid-liquid, and liquid-vapor interfaces, respectively. The relationship of Eq. (2.1) is known as the Young's equation and was tested later thermodynamically by Gibbs [1] and more recently by Johnson [2].

The wettability behavior of molten metals on a solid substrate has become the fundamental aspect in the manufacture of metal-ceramic composites, since in many manufacturing methods, the metal exists as a liquid phase in some stage of the process in contact with a solid ceramic. There are numerous reviews of articles on this subject, and one of the most complete was written by Naidich [3]. The degree of wettability is measured from the contact angle between the molten metal and the ceramic, as is shown in Fig. 2.2.

Wettability occurs when $\theta < 90°$. If the adhesion of the liquid molecules on the solid surface is weak compared to the cohesion between the molecules, the liquid will not wet the solid and the capillary level decreases as shown in Fig. 2.2a. For the case of Fig. 2.2b, there is a greater adhesion between the capillary wall and the liquid molecules, presenting a greater capillarity and of course good wettability.

© Springer Nature Switzerland AG 2018
A. Contreras Cuevas et al., *Metal Matrix Composites*,
https://doi.org/10.1007/978-3-319-91854-9_2

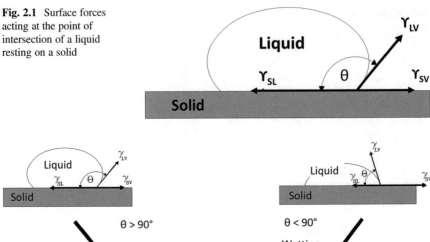

Fig. 2.1 Surface forces acting at the point of intersection of a liquid resting on a solid

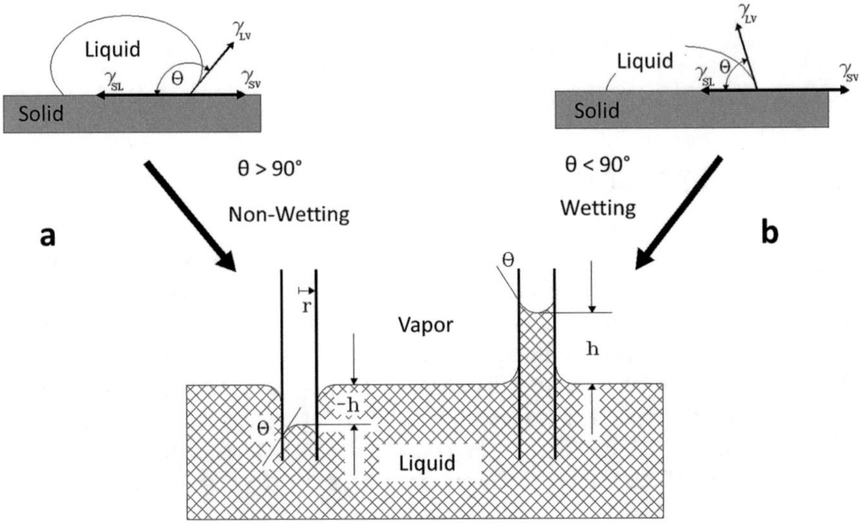

Fig. 2.2 (**a**) Non-wetting system. (**b**) Wetting system in function of the contact angle

2.2   Surface Tension and Adhesion Work

The surface tension of a liquid (γ_{LV}) in an inert gas atmosphere is a direct measurement of the interatomic forces acting on the surface. A related term is the *cohesion work*, which can be understood as the work done to divide a liquid column and create two new liquid/gas surfaces, such that:

$$W_c = 2\,\gamma_{LV} \tag{2.2}$$

The surface and interface terms are frequently considered by some authors as analogues, that is to say, with the same concept.

Equation (2.1) can be interpreted more widely considering the work of adhesion (W_a), which is defined as:

$$W_a = \gamma_{SV} + \gamma_{LV} - \gamma_{SL} \tag{2.3}$$

and is a useful measure that reflects the degree of bonding of solid and liquid surfaces [4]. If Eqs. (2.1) and (2.3) are combined, then:

$$W_a = \gamma_{LV} \left(1 + \text{Cos } \theta\right) \qquad (2.4)$$

Because both γ_{SL} and γ_{SV} are difficult to measure, Eq. (2.4) is more practical and both γ_{LV} and θ are measurable properties. Hence the adhesion work can be determined experimentally by evaluating the contact angle.

The condition for spontaneous wetting of the drop is given by:

$$W_a \geq 2\gamma_{SL} \qquad (2.5)$$

This means that the adhesion energy between the ceramic and the molten metal should be more than twice the surface tension of the liquid [5].

Equation (2.3) defining the adhesion work shows that the lower solid-liquid interfacial energy (γ_{SL}), the larger adhesion between the solid and the liquid; therefore there is a stronger bond between them [4, 6]. The mechanical strength of the bond after the solidification can be evaluated with some degree of approximation from W_a in the liquid state [7].

For a ceramic, it is said that the higher melting point tends to have higher surface energy; therefore, it is expected to have stronger bonds [8]. Table 2.1 shows the surface tension for some liquid metals at their melting point.

It is well known that the adhesion work between a ceramic and a metal decreases with the increase of the heat of formation of the carbides. This high heat of formation of the stable carbides implies a strong interatomic union and consequently a weak interaction with the metals, that is to say, a poor wet exists. Thus, highly ionic ceramics such as alumina are relatively difficult to wet because their electrons are tightly bound. Metal and covalent bonds are more similar in character, and covalent-bonded ceramics are more easily wetted by metals (and are more likely to react with metals) than highly ionic ceramics. The high valence electron concentration generally implies lower carbide stability and improved wettability between ceramics and metals. High temperatures and prolonged contact times usually promote chemical reactions that induce wetting.

Table 2.1 Surface tension for pure metals at melting temperature [9–11]

Metal	γ_{SL} (mJ/m^2)
Li	400
Mg	560
Zn	780
Al	870
Cu	1300
Ti	1650
Cr	1700
Ni	1780
Fe	1880
Mo	2250

2.2.1   Interactions that Determine the Work of Adhesion

The necessary condition for wetting in vacuum is $W_a > \gamma_{lv}$. The condition for spontaneous spreading during wetting is $W_a > 2\gamma_{LV}$. This means that a liquid metal will wet a solid only if the energy of the joints that are created through the interface exceeds the surface tension of the liquid.

When one considers how the molecules adhere to a solid surface, it can distinguish physical adsorption, which is governed by interactions of the van der Waals of chemical adsorption, which involves the formation of a chemical bond.

The contribution to the adhesion work due to a chemical reaction can be evaluated if the standard free energy change $\Delta G°$ can be estimated for the reaction. As a rule, when a chemical reaction occurs, its contribution to the work of adhesion is greatly imposed on the contribution by physical interactions. These reactions are also characterized by their high dependence on temperature, in contrast to physical interactions.

2.3   Wettability Systems

For a better understanding of wettability, we can classify it into two different types, (1) reactive systems and (2) nonreactive systems, depending on the reactivity and stability of the solid phase in contact with the liquid metal.

Nonreactive systems are generally characterized by extremely fast wetting kinetics, change in contact angle over a narrow range of temperatures, the nature of the solid surface is generally not modified by contact with the metal phase, and degree of wet is only the result of the establishment of the chemical equilibrium achieved in the bonds by the mutual saturation of the free valences of the surfaces in contact. The establishment of such bonds is not accompanied by the rupture of the interatomic joins in the present phases.

In reactive systems, the wetting of the ceramic phase by the liquid metal occurs with the extension of a chemical reaction and the formation of a new phase at the metal-solid interface. These types of systems are characterized by a remarkable wetting kinetics (important variations of contact angle over time) and a strong temperature dependence on contact angle changes [12].

When a drop of liquid metal is put into contact with a solid substrate, several phenomena involving mass transfer occur, such as the diffusion of atoms from one phase to another, adsorption, evaporation, or chemical reactions leading to the formation of new phases. This phenomenon continues until equilibrium is reached. The extent to which these reactions proceed depends on the nature of the substances involved in the wet as well as kinetic factors. However, sometimes it is difficult to distinguish strictly between reactive and nonreactive wetting. Some chemical interactions occur even in systems that are apparently not reactive. Reduction of a solid oxide wetted by a liquid metal is a typical case.

In summary we can establish the following:

1. Nonreactive systems:

   • Extremely rapid kinetics of wetting.
   • Very narrow temperature range to change contact angle.
   • The nature of the solid surface is not significantly modified by its contact with a metallic phase.
   • The degree of wetting is only the result of the establishment of chemical equilibrium bonds achieved by the mutual saturation of the free valences of the contacting surfaces.

2. Reactive systems:

   • Wetting occurs with extensive chemical reaction.
   • There are formations of new interfaces.
   • These systems are characterized by the pronounced kinetics of wetting (important variations of contact angle with time).
   • Strong temperature-dependent variations.

2.3.1 Classification of Wettability Systems

Based on the concepts of reactive wetting and nonreactive wetting, the wetting angles can be classified into two cases: (1) where the system is in equilibrium and (2) where the system is not in equilibrium. When there is equilibrium, the contact angle can be established from Young's equation in terms of static interfacial tensions as described above by Eq. (2.1). However, if the system is not in equilibrium, for example, a chemical reaction takes place between the two phases, then there is a mass transfer through the interface.

The contact angle obtained at equilibrium belongs to the final composition of the phases, which may be quite different from the initial phases at the time of the first contact between the liquid and the solid at the beginning of the experiment. In practice it is common to observe the change of contact angle over time and determine the period of time necessary for the contact angle to reach a constant value. Thus, the data found in the literature consist mainly of contact angles in the equilibrium. Some authors report values measured hours after the formation of the liquid drop, because even after long periods of time, the equilibrium has not been reached.

Some authors have investigated dynamic wetting by continuously measuring the contact angle throughout the experiment. However, these studies do not confirm the assumption proposed by Aksay [13], who proposes that the variation of the contact angle with time passes through a minimum before reaching the final equilibrium value in systems characterized by strong chemical reactions (Fig. 2.3).

Naidich [14] assumes that the liquid is chemically absorbed on the solid surface. The energy of the bonds formed between the two phases is the adhesion work as shown in Eq. (2.4). The bonding energy must then be equal to the energy of the

Fig. 2.3 Variation of interfacial tension γ_{SL} over time in non-equilibrium conditions: (1) adsorption, (2) reaction between the two phases

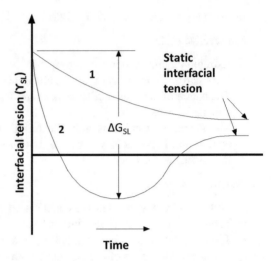

system. Thus, in the course of a chemical reaction, the released energy can be matched to the adhesion work.

Samsonov [15] explained the behavior of the wettability of metals with carbides based on the differences of the electronic structure in d orbitals. He concluded that the non-wetting results entirely from vacancies or complete filling in d orbitals of metals and that the electron share is a decisive factor in obtaining good wettability. For example, the electronic configuration of the IVA-VIA group shows that carbides have the characteristic of electron donors.

2.4   Techniques for Measuring Wettability

There are a great number of methods by which one can quantify the wettability of a solid by a liquid metal; the most suitable methods to measure wettability at high temperature are the techniques *sessile drop and Wilhelmy plate.* In the sessile drop test, direct observation of the liquid drop on solid substrate is followed until reach the equilibrium. From such observations, the actual contact angle that can be made of the geometrical aspects of the drop, such as shape, contact area, height, etc., is estimated. The main advantage of the sessile drop technique is that it is a relatively simple experiment to perform and give a direct measure of the contact angle reflecting the behavior of the system under investigation. However, its accuracy has been questioned since it lacks repeatability due to aspects such as contact angle hysteresis and surface adsorption surface effects as well as purity of materials.

Muscat et al. [16, 17] performed several tests using this technique. They investigated the wettability of TiC by aluminum where the results obtained indicate that the contact angle seems to never reach a stable value, indicating a dynamic behavior where the contact angle changes with temperature and time.

In the Wilhelmy plate technique, sometimes also known as the plate weight method, a dish suspended from a scale is partially immersed in a liquid. The plate experiences an additional force resulting from the surface tension of the liquid at the liquid-solid-vapor interface. This force will be equal to:

$$F = \gamma_{LV} \, X \, \text{Cos} \, \theta \tag{2.6}$$

where X is the wetting perimeter, θ the contact angle, and γ_{LV} the liquid-vapor surface tension. The value of the contact angle (θ) will dictate the magnitude of the downward force due to the surface tension given by the above equation.

If we consider a bar immersed in a molten metal like the one shown in Fig. 2.4 similar to Wilhelmy plate used to measure contact angles, we will initially have a contact angle >90° (Fig. 2.4a). Over time and at a sufficiently high temperature, this force will be reversed and contact angle will be <90° (Fig. 2.4b).

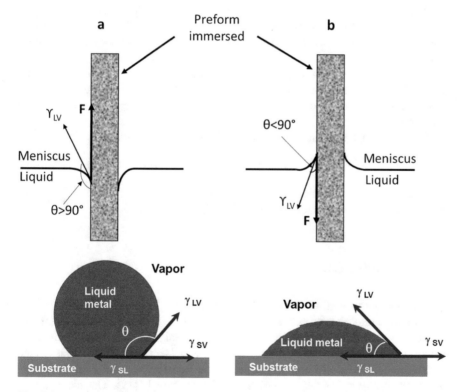

Fig. 2.4 Transitory effect of contact angle and resulting force due to surface tension, (**a**) for non-wetting system, (**b**) for a wetting system

2.5   Techniques to Improve Wettability

Many of the problems encountered in the manufacture of metal-ceramic composites are a consequence of the interface characteristics. The wettability of ceramic rein-forcement by molten metals and its alloys is one of the most important parameters to determine interfacial properties. In many cases, the wettability is poor; therefore suitable engineering methods have been implemented in the manufacture of com-posites. An efficient approach is to apply coatings to the ceramic reinforcement that reacts with the oxide layer. Another method is the addition of alloying elements that reduce the surface tension and improve wettability.

Wetting is favored by the formation of strong chemical bonds at the interface. Also, good wetting often involves good bond strength at the interface in the final compound. If the bond at interface is weak, it can be detrimental in the compound since this brings with it a decrease of the mechanical properties. Therefore, a balance between good wetting and a minimal interfacial reaction is required.

According to Young's equation (Eq. 2.1), the contact angle of a liquid with a solid can be decreased by (a) increasing the surface energy of the solid (γ_{sv}), (b) decreasing the interfacial energy between the solid and liquid (γ_{SL}), and (c) decreasing the surface tension of the liquid (γ_{LV}) [18]. The condition for a liquid to be spread on a solid is given by ($\gamma_{sv} > \gamma_{sl} + \gamma_{lv} \cos \theta$), meaning that low values of the surface tension of a liquid (γ_{lv}) are favorable for wetting; although a stable solid-liquid interface is formed, the adhesion work must also be high.

In addition, the wettability and bonding work between liquid metals and nonme-tallic solids can be increased by inducing an intense chemical reaction at the interface boundaries, the intensity of which depends on the loss of energy at the solid-liquid interface.

In summary we can said that contact angle (θ) can be decreased by:

- Increasing the surface energy of the solid, γ_{SV} ($\gamma_{SV} > \gamma_{SL} + \gamma_{LV} \cos \theta$), so spread can occur.
- Decreasing the interfacial energy γ_{SL}.
- Decreasing the surface tension of the liquid, γ_{LV}.

2.5.1   Metallic Coatings

To improve the wetting of ceramics by liquid metals, the application of a metallic layer to the ceramic has been used, which essentially increases the surface energy of the solid (γ_{sv}), promoting wetting with the liquid metal [18]. In this way, infiltration is more easily accomplished by the deposition of a metal coating on the surface of the reinforcement. The most frequently used metal as a coating is nickel. Nickel coatings are especially used for Al/SiC and Al/Al$_2$O$_3$ compounds [19]. The nickel reacts strongly with the aluminum to form stable intermetallic compounds (NiAl$_3$, Ni$_2$Al$_3$, etc.); in this way, the wetting is excellent. However, the embrittlement of these compounds is detrimental for the mechanical properties of the compound. Silver, copper, and chromium have also been used like coating in ceramics for the same purpose. The high solubility of silver in aluminum provides good wetting without involving the formation of brittle compounds [20].

2.5.2 The Addition of Alloying Elements

The most commonly used technique to promote wettability is the addition of an alloying element in the metal bath. The admission of an alloying element in the liquid metal can promote the wettability of a solid surface by three mechanisms: (1) reducing the surface tension of the liquid, (2) reducing the solid-liquid interfacial energy, and (3) by chemical reactions at the solid-liquid interface. Figure 2.5 shows the surface tension of aluminum in function of the percentage of alloying elements added at the temperature of 700–740 °C [21].

The effect of Mg on the reduction of Al surface tension is very noticeable as shown in Fig. 2.5. Therefore, a reduction in the contact angle is expected as the Mg content increases. However, it may be that some interfacial reaction is the one that affects in greater proportion the effect of this element on the wettability of Al-Mg alloys, since it is well known that alloys containing Al react at the interface with carbon to form aluminum carbide (Al_4C_3).

On the other hand, Mg is a highly reactive element in the presence of oxygen as shown the following reaction:

Fig. 2.5 Effect of alloying elements on aluminum surface tension at 700–740 °C [21]

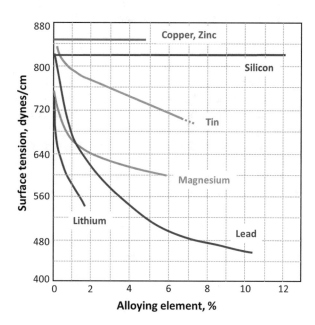

$$Mg_{(l)} + 1/2O_{2(g)} = MgO_{(S)}$$
$$\Delta G^{\circ}_{(900^{\circ}C)} = -473.19\,kJ \tag{2.7}$$

Mg may form magnesium oxide (MgO) or spinel ($MgAl_2O_4$) which have been reported to improve wettability [22], as well as the Mg can reduce the aluminum oxide according to the following reaction:

$$3Mg_{(l)} + Al_2O_{3(S)} = 3MgO_{(S)} + 2Al_{(S)}$$
$$\Delta G^{\circ}_{(900^{\circ}C)} = -123.86\,kJ \tag{2.8}$$

According to the thermodynamic considerations, the reduction of Al_2O_3 by the Mg is possible. If this reaction is carried out, the MgO may be the obstacle at the interface that prevents a real contact between molten Al and solid substrate.

Also Mcevoy et al. [23] report the formation of $MgAl_2O_4$ spinel in the Al/MgO system in a temperature range of 1000–1350 K and propose the formation of $Al_2O_{(g)}$ according to the following reaction:

$$2Al_{(l)} + MgO_{(S)} = Mg_{(S)} + Al_2O_{(g)}$$
$$\Delta G^{\circ}_{(900^{\circ}C)} = +236.26\,kJ \tag{2.9}$$

Thermodynamically this reaction is not possible. However, they indicate that entropically it is favorable. The formation of $Al_2O_{(g)}$ is reported by several researchers [23–26] as the main factor to break the alumina layer that forms in Al drop and to have a real Al/TiC interface to carry out the wet. If this latter fact occurs, according to the Ellingham diagram [27], Mg forms an oxide (MgO) which is more stable than Al_2O_3, and the formation of a magnesium oxide layer becomes the obstacle in the interface which prevents direct contact between the aluminum and TiC. López et al. [28] carried out a study of the influence of the alloying elements of commercial alloys (2024, 6061, and 7075) and of the atmosphere in the nature of the alumina layer where it proposes a mechanism for the breaking of this layer of alumina. Similarly, Madeleno et al. [29] propose a mechanism for breaking the oxide layer in alloy 6061 under vacuum.

Mg improves wettability more effectively under vacuum conditions due to its rapid evaporation. It is reported that during wetting experiments with Al alloys with 2.5% Mg on alumina, Mg volatilizes completely to 700 °C after 3 min of melting, which improves wettability [30]. Pech-Canul et al. [31, 32] propose a series of chemical reactions in the Al-Mg-N system to reincorporate the Mg loss by evaporation to Al. They suggest that wetting tests should be performed under a nitrogen atmosphere; thus this reincorporation is carried out.

According to the literature, the addition of magnesium to aluminum decreases surface tension as the magnesium content increases [21, 33, 34], while elements such as zinc, copper, manganese, and silicon seems to have little effect (Fig. 2.5). Narciso et al. [35] studied the effect of Mg and Si individually and in combination, on Al wettability over SiC substrates. They found that elements such as Mg and Si added to aluminum individually seems to have no effect on the contact angle in Al/SiC systems.

Additions of suitable alloying elements in liquid metals can reduce the surface tension of the liquid, which means a reduction in γ_{LV}. Elements of active surface usually accumulate in the interface, decreasing the interfacial energy. Therefore, the adsorption of an alloying element having a lower liquid/vapor interfacial tension (γ_{LV}) than the matrix liquid will cause a decrease in the solid-liquid interfacial energy (γ_{SL}) promoting wettability according to Young's equation. The addition of alloying elements to the baths to promote wettability between the liquid (matrix) and the solid phase (reinforcement) has been widely used by various researchers [18, 36, 37].

Magnesium present in the bath (regardless of the addition form) can be enriched at the surface of the dispersant according to the Gibbs adsorption equation and under favorable conditions can also react with the oxides to form a reaction product at the interface [38].

The addition of an alloying element to the molten metal can influence the adhesion work (W_a) and directly the wetting angle θ, by absorbing the alloying element in the metal-ceramic interface or the metal/atmosphere interface leading to a decrease in surface tensions γ_{ls} and γ_{lv} [39]. The interfacial adsorption of alloying elements improves both adhesion and wettability. However, surface adsorption always deteriorates adhesion and only improves wettability if θ is originally below 90°.

2.5.3 Modification of the Reinforcement Phase

Another alternative that has been studied and used for the purpose of improving wetting has been the modification of the ceramic phase, either by the stoichiometric variation of the compound or by some heat treatment that superficially modifies the reinforcing phase.

2.5.4 The Use of Ultrasound

It has been reported that the use of ultrasonic vibrations improves the wettability of the ceramic particles with the molten metals, possibly due to the partial desorption of the gases absorbed from the surface of the ceramic particles [40].

2.5.5 Heat Treatment of the Reinforcement Phase

The investigation focused in the study of wetting with reinforcement surface modification has been carried out mainly in the SiC/Al system. Laurent et al. [41] studied the wettability of SiO_2 and oxidized SiC with aluminum, finding that the strong reactivity between Al and SiO_2 cannot be used to improve the wetting of this metal on SiC. Bardal [42] studied heat-treated SiC particles at 1100 °C producing an oxide layer on the surface of the particles and by means of an infiltration technique studied the wettability of these oxidized particles and SiC with an $AlSi_6Mg$ alloy. He found that SiC oxidized particles are wetted more easily than SiC particles. The studies indicate that Mg spinels formed in this system are wetted more easily by aluminum. Thus, aluminum alloys containing Mg improve wetting on oxidized surfaces through formation of spinels in the interface.

2.6   Oxygen Influence

In all metal-ceramic systems, oxygen has a particular interest, since this element is present in most processes and because it has influence on the surface properties of many metals at partial pressures as low as 10^{-15} atmospheres [43, 44]. The interaction of liquid metals with oxygen results in oxide formation. A typical example of this is the case of aluminum. Near to its melting point, the surface of liquid aluminum is covered by a small layer of aluminum oxide that inhibits the formation of Al-ceramic interfaces. The use of static atmospheres of inert gases retards evaporation of the oxide layer and displaces this transition at high temperatures [45], while additions of Mg or Ca to the molten aluminum interact with the oxide layer to reduce its effect.

Oxygen dissolution increases the wettability of oxides by liquid metals, although the oxygen concentration may be as low as a few ppm or tenths of ppm. An explanation of this phenomenon was given by Naidich [3] who proposed that the solution oxygen in the metal associated with atoms of the metal forms groups that have a partially ionic character, which results from the transfer of charge of the metal to the atoms of oxygen. These groups can develop Coulomb interactions with ionic-covalent ceramics and, consequently, strongly adsorb the oxide-metal interface. It is assumed that the adhesion with the oxides is essentially governed by the interactions of the metal atoms with the oxygen anions.

Generally, the oxides wettability by pure metals is poor. Humenik [46] found that the work of adhesion of the liquid metals on oxidized surfaces increases with the increase of the affinity of the metal by the oxygen. In the case of binary alloys, this is manifested by a strong adsorption at the interface of the most electropositive metal. In the same way that W_a increases with increasing affinity of the metal for oxygen, W_a also increases with decreasing strength of the bonds between the metal and oxygen in the solid. This increase in the interaction is frequently correlated with the dissolution of the oxide in the molten metal bath, since the W_a increases drastically with the increase of the oxygen concentration in the bath. The adhesion work becomes approximately equal to the energy required to break the (ionic) bonds between the two oxides.

2.7   Effect of Roughness on the Contact Angle

The first attempts to predict the effect of roughness on the apparent contact angle ϕ were made by Wenzel [47], who proposed the relation:

$$\text{Cos } \phi = r \cos \theta \tag{2.10}$$

where r is the roughness factor and is defined as the ratio of the real area A_r of the solid to the apparent area (A) of a plane, having the same macroscopic dimensions. The θ angle is the intrinsic contact angle defined by the Young equation. This equation is important because it explains a fundamental observation: a rough surface improves wetting if $\theta < 90°$, whereas this leads to a greater apparent contact angle if $\theta > 90°$. This equation has also been deduced from thermodynamic fundamentals.

Nakae et al. [48] conducted an extensive study on effects of roughness height on wettability using the Wenzel roughness factor. Although it has not been fully understood, the hysteresis effects are mainly caused by a rough surface, a heterogeneous surface, impurities that are absorbed at the surface, and rearrangement or alteration of the surface by a solvent. The main effects that a rough surface produces at the contact angle is to have receding contact angles; thus care must be taken when making these measurements; otherwise you would have erroneous measurements.

2.8   Spreading Kinetic

Drop spreading can be divided into different stages. The first stage is characterized by a rapid spreading of the drop because it tends to achieve a balance of the interfacial tensions characteristic of nonreactive liquids. In the second stage, the liquid dissolves the solid substrate to form interfacial compounds.

The dynamics of contact angle depend on the rate of drop spreading, and several relationships between contact angle and velocity have been proposed in the literature [49–56].

A kinetic model is applicable in a given system when the drop spreading is plotted as ln r vs ln t adjusted to a straight line, where r is the instantaneous radius of the drop and t is the time. Each system shows different behavior in the drop spreading, and some follow a linear relation as $(r \sim t)$ or $(r^4 \sim t)$ which is characteristic of reaction mechanisms and controlled diffusion. However, the kinetics of drop spreading of a given system is strongly affected by the experimental conditions, being able to obtain very different kinetics of drop spreading for the same system [57, 58].

In nonreactive systems the drop spreading rate is controlled by the viscous flow because the viscosity of the molten metal is very low, and the time required for a millimetric drop to reach the capillary equilibrium in a nonreactive system is less than 10^{-1} s. This time is several times shorter than the observed spreading time in metal-ceramic reactive systems, which are in the range of 10^1–10^4 s. Therefore, in reactive systems, the velocity of propagation is not controlled by viscous resistance, but it is controlled by interfacial reactions [9, 49, 13].

2.8.1 Ridges Effects

Ridges can form at the triple line in response to the vertical component of force from surface tension. Measurements of contact angle are interpreted in terms of the Young-Dupre equation, which establishes that a liquid in thermodynamic equilibrium with a rigid solid will assume a characteristic macroscopic contact angle at the triple junction, which depends on the surface energies. When the substrate is not perfectly rigid and inert, the vertical force causes deformation of the solid at the triple point. In these cases, the vertical component of surface tension is resisted by elastic distortions of the solid. In high-temperature systems (e.g., molten metals on ceramics), the experimental temperatures are typically between 0.2 and $0.5 T_m$ (T_m = the substrate melting temperature), which will lead to some local diffusion or solution precipitation. Under these conditions, a small ridge will eventually develop at the triple junction as is shown in Fig. 2.6.

It is proposed that during spreading, a small enough ridge can be carried by the triple junction under certain conditions, resulting in a variable macroscopic angle as is shown in Fig. 2.7. As a consequence, motion of the triple line can be inhibited by drag of the ridge, yielding spreading rates that are orders of magnitude slower than for viscous drag controlled flow.

The spreading process can be divided into four stages depending on the degree of evolution of the triple point ridge [59]. The first corresponds to the classical, rigid (elastic) solid and occurs during short times wherein liquids spread rapidly at high driving force and a triple line ridge would be unstable or at such low temperatures that only elastic ridges can form. Second stage prevails under conditions in which a ridge exists that allows two-dimensional equilibrium to occur at the junction, and yet is small enough that it can be carried by the liquid front, and, thereby, allow establishment of a macroscopic contact angle that virtually obeys the Young-Dupre equation. In the third stage, a ridge grows rapidly compared with the spreading velocity. The most obvious cause for this is if the ridge is becoming big compared to the radius of curvature and the solid is becoming appreciably distorted. If spreading occurs in this regime, the measured contact angles cannot be interpreted in terms of Young's equation. Finally, at longer times more extensive deformation of the solid will arrive and the system will be in another transitional phase. Stage IV corresponds to complete equilibrium at the triple point and constant curvature shapes.

Fig. 2.6 Geometry of the
drop in function of time,
(**a**) initially (**b**) during the
spreading kinetic a ridge
will form in the triple point,
(**c**) diffusion of the liquid
drop through substrate

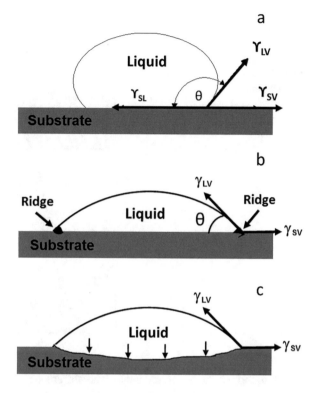

2.8.2 Advancing and Receding Contact Angles

Wetting is important in the bonding or adherence of two materials. Wetting and the
surface forces that control wetting are also responsible for other related effects,
including capillary effects. The degree of wetting (wettability) is determined by a
force balance between adhesive and cohesive forces. Wetting deals with the three
phases of materials: gas, liquid, and solid. The contact angle (θ), as was shown in
Fig. 2.1, is the angle at which the liquid-vapor interface meets the solid-liquid
interface. The contact angle is determined by the result between adhesive and
cohesive forces. As the tendency of a drop to spread out over a flat, solid surface
increases, the contact angle decreases. Thus, the contact angle provides an inverse
measure of wettability.

The derivation of Young's equation assumes that the solid surface is smooth,
homogeneous, and rigid; it should also be chemically and physically inert with
respect to the liquids to be employed. Ideally, according to Young's equation, a
unique contact angle is expected for a given system. In a real system, however, a
range of contact angles is usually obtained. The upper limit of the range is the
advancing contact angle, θ_a, which is the contact angle found at the advancing edge
of a liquid drop. The lower limit is the receding contact angle, θ_r, which is the contact

Fig. 2.7 Profiles of a ridge
that will form by local
diffusion in response to
horizontal and vertical
forces in the triple
junction [59]

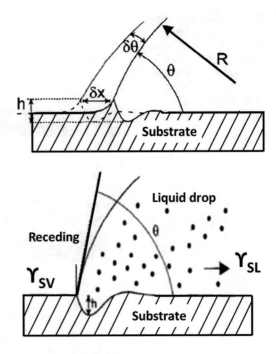

angle found at the receding edge. The difference between the advancing and
receding contact angles is known as the contact angle hysteresis (θ_{hyst}) [60]:

$$\theta_{\mathrm{hyst}} = \theta_a - \theta_r \tag{2.11}$$

Practically, all solid surfaces exhibit contact angle hysteresis. However, this
hysteresis interpretation in terms of Young's equation is complicated. Not all the
experimentally measured or observed contact angles are reliable and appropriate.
Although contact angle hysteresis has been studied extensively in the past several
decades, the underlying causes and its origins are not completely understood.
Studies have attributed contact angle hysteresis to surface roughness [61–63] and
heterogeneity [64–66], as well as metastable surface energetic states [64, 65]. Some
found that the hysteresis decreases with increasing molecular volume of the liquid on
monolayers [67]. In more recent studies, contact angle hysteresis was found to be
related to molecular mobility and packing of the surface [68, 69], liquid penetration,
and surface swelling [70]. Previous studies [60, 71] showed that contact angle
hysteresis is strongly dependent on the liquid molecular size and solid-liquid contact
time. These findings lead to the presumption that liquid sorption and liquid retention
are causes of contact angle hysteresis.

2.9 Wettability of TiC by Aluminum

The wettability behavior of molten metals on a solid substrate has become a key aspect in the manufacture of composites. The wettability of TiC by molten aluminum has been extensively investigated. However, the temperature at which molten aluminum wets TiC has not been precisely established, due to there are reported different temperatures which change due to the experimental conditions used. Jha et al. [72] argue that liquid aluminum wets TiC above 700 °C ($\theta = 118°$).

The first studies on wettability of TiC by aluminum were performed by Rhee [73] and Kononenko [74] using the sessile drop technique. As shown in Fig. 2.8a, Rhee [73] evaluated wettability in the TiC/Al and AlN/Al system at temperatures between 700 and 900 °C and did not report interfacial interactions. Good wetting was obtained for TiC/Al system around 750 °C. On the other hand, Kononenko [74]

Fig. 2.8 (a) Change of contact angle from 700 to 900 °C for Al/AlN and Al/TiC systems [73], (b) wettability of TiC by pure aluminum from 860 to 960 °C [16]

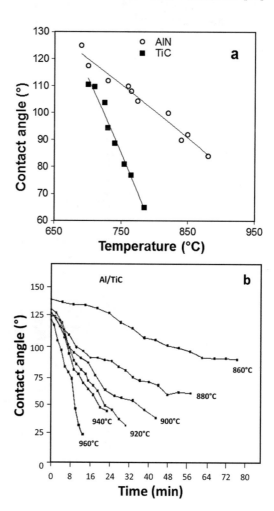

reports a transition from non-wetting to wetting at about 1050 °C, while Samsonov [15] reports an angle of 148° at 1100 °C, 90° after 10 min at 1150 °C, and 60° after 5 min at 1200 °C.

Probably the most complete studies have been performed by Muscat et al. [16]. Figure 2.8b shows the wetting kinetics of TiC by pure aluminum in the temperature range of 860–960 °C. The instability of the contact angle at the different temperatures was observed, with the faster decrease of the contact angle when increasing the temperature, indicating a dynamic behavior and good wettability in the system at temperatures above 860 °C. The difference between results of different researcher could be attributed to the controlled conditions during the experiments, mainly due to the inert gas or vacuum used.

The increase in wettability in the Al/TiC_x system is attributed to the changes in the nature of TiC bonds. In particular at low values of x, the metallic character of its bonds is more dominant. Interaction with liquid aluminum and eventual dissolution of the carbide together with the increase of Ti content at the interface may prevent the formation of Al_4C_3 [75]. Contact angle for some ceramic/metal systems are presented in Table 2.2.

Table 2.2 Wetting angles for some ceramic/metal systems

Metal	Ceramic	Temperature (°C)	Angle	Reference
Si	Si_3N_4	1500	40°	[12]
Si	Graphite	1450	0°	[76]
Cu	Graphite	1100	157°	[76]
Cu	Graphite	1135	144°	[77]
Cu	TiB_2	1135	139°	[77]
Cu	TiC	1135	130°	[77]
Cu	SiO_2	1135	131°	[77]
Cu	TiO	1150	75°	[12]
Al	TiB_2	700	92°	[76]
Al	TiB_2	900	37°	[76]
Al	Graphite	700	145°	[12]
Al	Graphite	1100	57°	[12]
Al	AlN	700	160°	[12]
Al	AlN	1100	50°	[12]
Al	Si_3N_4	1100	60°	[12]
Al	ZrO_2	1100	87°	[12]
Al	SiO_2	700	150°	[12]
Al	SiO_2	1100	50°	[12]
Al	TiC	700	118°	[76]
Al	TiC	900	59°	[78]
Al	TiC	1100	10°	[77]
Al	Al_4C_3	1100	60°	[79]

2.9.1 Stoichiometry Variation of the Ceramic

The wettability of the ceramic surfaces by the liquid metals depends strongly on the stoichiometry of the ceramic compound, either an oxide or a carbide. Recently, Frumin et al. [75] studied the wettability of the stoichiometric and substoichiometric TiC_x for $0.5 < x < 1$ with pure aluminum in the temperature range of 700–1100 °C. They also observed, as in other systems, a transition from non-wetting to wetting, which depended on the stoichiometric composition of TiC; at a lower value of x, the transition begins at lower temperatures. This effect is shown in Fig. 2.9.

They point out that non-wetting at temperatures below 900 °C is due to the presence and stability of aluminum oxide layer covering the liquid drop. In addition they indicate that at a lower value of x, the metallic character of the bond in the TiC is more dominant, leading to a greater chemical interaction with the Al. They also conclude that only TiC_x with $x < 0.90$ coexists in equilibrium with the aluminum at relatively high temperatures (>900 °C). For high values of x, the TiC becomes unstable and aluminum carbide is formed.

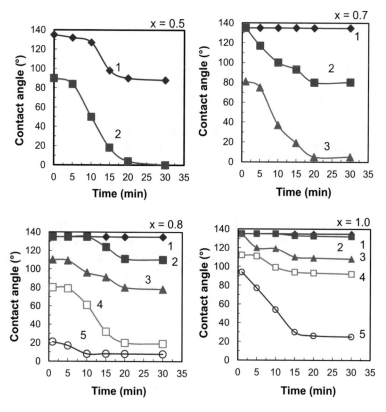

Fig. 2.9 Time dependence of the contact angle for TiC_x with pure Al for (1) 700 °C, (2) 800 °C, (3) 900 °C, (4) 1000 °C, and (5) 1100 °C [75]

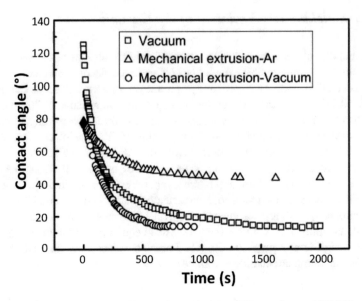

Fig. 2.10 Change of contact angle with time for molten Al on $TiC_{0.7}$ surfaces at 1000 °C in vacuum and argon atmosphere using improved and modified sessile drop method [80]

More recently, Lin et al. [80] studied the wetting of titanium carbide ($TiC_{0.7}$) by liquid Al, using the sessile drop method at 850–1050 °C. An improved sessile drop technique which prevented the oxidation of aluminum was used to measure the changes in contact angle.

The results show that the aluminum surface oxidation and the thickness of the oxide film have a pronounced effect on the wettability, especially at low temperatures. To eliminate this effect, the experimental temperature must be over a critical value. Vacuum favors lowering this value compared with atmosphere, and the improved sessile-drop method, particularly using an impingement-dropping mode, helps to weaken this effect by mechanical disruption and removal of the oxide film.

Figure 2.10 shows the variation in the contact angle with time for molten Al on the $TiC_{0.7}$ substrates at 1000 °C in vacuum and Ar atmosphere using modified and improved sessile drop methods. Considerable differences could be observed. Firstly, the initial contact angle is much smaller for the extruded drop obtained by the improved method (78°) than that for the non-extruded method (125°). Clearly, the smaller contact angle obtained by the modified method is influenced by the presence of the oxide film on the Al drop surface. Secondly, the final equilibrium contact angle reached in the Ar atmosphere is 43° while that in vacuum is 14° for the non-extruded drop and 12° for the extruded drop, indicating that the high vacuum favors a smaller oxygen partial pressure and thus a clean drop surface. The interfacial reaction yielded Al_4C_3. The wetting is essentially improved initially by the formation of the Al_4C_3 phase and then by the adsorption of Ti at the solid-liquid interface. The dissolution of $TiC_{0.7}$ and reaction with molten Al lead not only to the formation of the Al_4C_3 and Al_3Ti phases at the interface but also to the increases in the stoichiometry of TiC and the amount of the Ti content in the Al drop.

2.9.2   Interfacial Reactions

The nature of atomic bonds and the thermodynamic stability of solids in contact with liquids are related to wettability at high temperatures. In the case of carbides, the wettability decreases with increasing the formation heat of the carbide. The high heat of formation implies strong interatomic connections and weak interaction with metals, which is reflected in poor wettability. In the case of oxides, wettability decreases by increasing the free energy of oxide formation. Highly ionic ceramics such as alumina are hardly wet. Ceramics with covalent bonds are more easily wetted by metals (more prone to react) than ceramic with ionic bonds, due to the similarity between the metallic and covalent bond. High temperatures and prolonged contact times generally promote wettability through a chemical reaction [57, 81].

The wetting and spreading behavior of reactive systems is further complicated by interfacial reactions at the solid-liquid interface as well as reactions at the liquid-vapor interface (e.g., oxidation).

From thermodynamic data it can be concluded that TiC_{1-x} and aluminum react to form Al_4C_3 and $TiAl_3$ at temperatures below 800 °C. TiC_{1-x} is in equilibrium with Al_4C_3, $Al_{(l)}$, $TiAl_3$, and TiAl, but not in equilibrium with Ti_3Al at 1000 °C [82]. Due to the particular importance of aluminum carbide in the manufacture of composites, Iseki et al. [83] performed a study on some of the most important properties of this compound.

Some years ago, Banerji [84] reported interesting studies about the process to form TiC particles, reacting Ti and C in the molten Al. They proposed that Ti reacts with C at temperatures below 1000 °C, forming TiC according to the following reactions:

$$Ti + C = TiC \tag{2.12}$$

$$TiC + 3Al = Al_3Ti + C \tag{2.13}$$

And the liquid aluminum available for wetting TiC should induce the following reactions at the particle/metal interface:

$$3TiC + 4Al = Al_4C_3 + 3Ti$$
$$\Delta G^\circ_{973\,K} = +350\,kJ/mol \tag{2.14}$$

$$9Ti + Al_4C_3 = 3Ti_3AlC + Al \tag{2.15}$$

The relative stabilities of Al_4C_3 and TiC in the molten aluminum bath will depend on their respective ΔG formation. Fine and Conley [85] suggested that it is necessary to compare the free formation energy of Al_4C_3 and TiC on the basis of one mole of C.

TiC has a lower free formation energy than Al_4C_3 over a temperature range up to 1800 °C. Therefore, it would be expected that reaction (2.14) should favor the formation of TiC. Therefore, they indicate that the explanation of Banerji and Reif

[84] to the presence of Al_4C_3 and Ti_3AlC in their master alloys below 1000 °C must come from another source.

Fine and Conley [85] propose that the aluminum carbide observed is formed for kinetic reasons and disappears by reaction (2.15). Of course, this reaction will be faster at higher temperatures. They also suggest that the Ti_3AlC phase with the perovskite structure is formed according to the reaction:

$$3Al_3Ti + C = Ti_3AlC + 8Al \qquad (2.16)$$

And it is present below 1000 °C as a metastable phase, intermediate between $TiAl_3$ and TiC. The TiC is unstable in the aluminum bath below 1450 K (1177 °C), which favors the reaction (2.14), followed by the reaction (2.15) at the Al/TiC interface. This will lead to the formation of the compounds Al_4C_3 and Ti_3AlC, surrounding the TiC particles.

On the other hand, Yokokawa [86] considering the chemical potentials in the Al-Ti-C system indicate that $Al_{(l)}$, TiC and $TiAl_3$ can coexist in equilibrium at 1000 °C. However, at 700 °C, $Al_{(l)}$ does not have a contact in equilibrium with TiC. This characteristic can be interpreted in terms of the following reaction:

$$3TiC + 13Al_{(l)} = Al_4C_3 + 3TiAl_3$$
$$\Delta G^\circ_{973\,K} = -16.86\,kJ/mol \qquad (2.17)$$

The Gibbs free energy change for this reaction is -16.86 kJ/mol at 700 °C and 55.95 kJ/mol at 1000 °C. From this consideration, the results of Banerji and Reif on the formation of Al_4C_3 surrounding TiC particles can be interpreted as originating in equilibrium. As can be seen, there is great controversy regarding the stability of TiC in the presence of liquid Al.

2.10   Wettability of TiC by Pure Al and Pure Mg

TiC is one of the ceramics more used in the fabrication of composites that was the reason that wetting of TiC was studied with two of the more used metal (aluminum and magnesium) in fabrication of MMC. The wetting behavior of TiC by liquid aluminum and magnesium was studied by the sessile drop method in the temperature range of 800–1000 °C [87, 88].

The wetting of ceramic surfaces by molten metals is one of the most important phenomena to consider when producing a metal matrix composite material, especially by processes involving liquid metals. Many of the problems encountered in the fabrication of aluminum/ceramic composites are a consequence of the characteristics of wettability.

The wetting of ceramic surfaces by liquid metals has been the subject of study, and it is well known that in many systems the wetting process depends on the chemical reaction occurring at the solid-liquid interface. However, ceramic materials are frequently not wetted by liquid metals. The basic reason is that most of the ceramics are ionic or covalent in nature and are not compatible with the metallic species.

Most of the work on MMC was based in light metals such as Al, because of its high electrical conductivity and low density. Recently, Mg has become an important metal to use as a matrix in the fabrication of MMC since Mg is a very light metal that is approximately two-thirds as dense as Al and offers distinct advantages over aluminum. Mg does not form a stable carbide, so carbides like TiC should be suitable reinforcements for Mg matrix composites.

In most cases in ceramic/metal systems, the transition from non-wetting behavior to a wetting behavior is between 900 and 1000 °C [16, 76–78]. However, good wetting at 800 °C was observed [87].

2.10.1 Experimental Conditions

Titanium carbide powder (H. C. Starck, grade c.a.s.) with an average particle size of 1.2 µm was used to fabricate dense TiC substrates with 96% average theoretical density. Figure 2.11a show a SEM image of TiC powders and Fig. 2.11b show the frequency of particle size of TiC powders. The TiC substrates were prepared in a 25 mm diameter graphite die at 1800 °C for 30 min in a vacuum furnace with a graphite heater with a pressure applied of 30 MPa. Figure 2.12a, b shows an image of the hot-press used and Fig. 2.12c, d an image of the surface of the TiC substrate and measurement of roughness through atomic force microscopy (AFM), respectively. Values of roughness are between 2.6 and 2.8 nm.

The substrates were polished on one face using 1 µm diamond paste for the final polish and cleaned ultrasonically in acetone. Pure Al and Mg samples were cut in small cubes of ~0.5 g, which were polished and cleaned in similar way of the substrates in order to reduce at the maximum the oxidation.

Fig. 2.11 (a) SEM micrograph of as-received TiC powders and (b) frequency of particle size

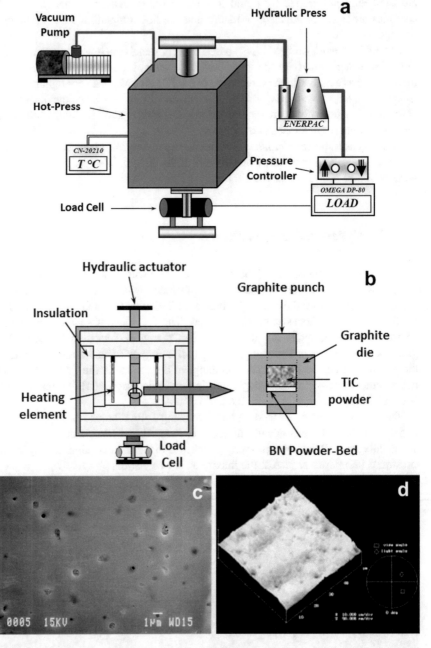

Fig. 2.12 (**a**) Experimental setup of the hot-press used, (**b**) schematic view of internal chamber of hot-press, (**c**) polished surface of substrates, (**d**) measurement of roughness

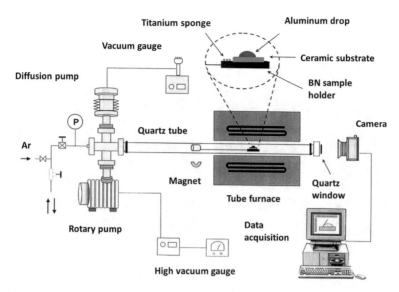

Fig. 2.13 Experimental setup used to carry out the wetting tests (sessile drop)

The purity of the aluminum employed was 99.99% and electrolytic Mg was used to carry out the wettability tests in the temperature range of 800–1000 °C under a static inert atmosphere of argon, using the sessile drop technique. The wetting experiments were conducted within a sealed quartz tube heated by a resistance furnace, under a protective atmosphere of Ar (99.99%) to prevent oxidation of the melt. An schematic representation of experimental setup is shown in Fig. 2.13.

2.10.2   Wetting and Spreading Kinetic

At the time the metal drop was well shaped, the contact angle and the drop base radius were recorded photographically at various time intervals followed until a steady state was reached. Figure 2.14 shows several images of the melting drop on the substrate several times. Image processing software was used to measure the contact angle, the height, and diameter of the drops in order to evaluate the kinetic spreading. Measures in the left and the right side of the drop were carried out.

After finish the experiment, the samples were immediately removed from the hot zone inducing fast cooling. The solidified drops were sectioned and polished to examine possible interfacial reactions using scanning electron microscopy (SEM), energy-dispersive analysis (EDS), and electron probe microanalysis (EPMA).

Typically 120 min were necessary to reach a stable contact angle in the Al/TiC system. In the case of pure Mg/TiC system, the experiments were shorter, requiring around 30 min to obtain stable contact angle, and after this time the shape of the drop was lost due to the evaporation of Mg and the top part of the drop started

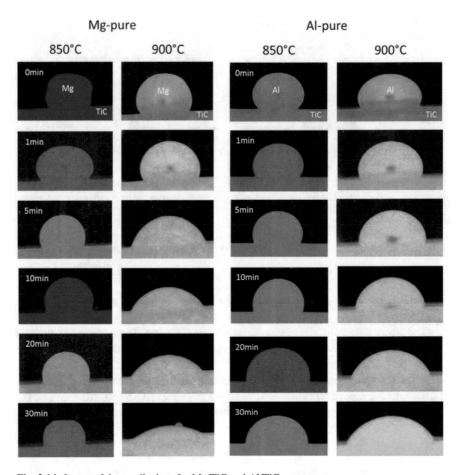

Fig. 2.14 Image of the sessile drop for Mg/TiC and Al/TiC systems

to flatter. To stabilize the shape of the drop, an extra Mg ingot was placed in the hot zone prior to introduce the sample.

Figure 2.15a shows contact angles and Fig. 2.15b the drop base radius for the Al/TiC system as a function of time and temperature. These curves exhibit three regions of the contact angle with respect to time: the first typified by a slighter slope (deoxidation of the Al drop), the second showing a sharp slope (chemical reaction), and the third where the value of θ is nearly constant. More than 120 min were generally required to obtain a constant angle. Good wettability was observed in all the temperatures used. The spreading radius increased with time in similar way like the contact angle decrease with the time until reached equilibrium.

When the molten aluminum is in real contact with the surface of TiC, spreading is driven solely by chemical reaction at the interface. The formation of the gaseous compound Al_2O is considered to play a dominant role in breaking up the alumina film surrounding the molten aluminum drop that leads to a decrease in contact angle.

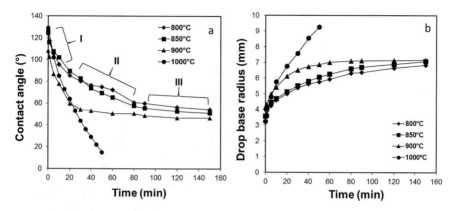

Fig. 2.15 (**a**) Variation of the contact angle and (**b**) the drop base radius with time for the Al/TiC system [87, 88]

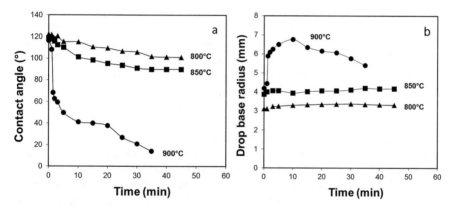

Fig. 2.16 (**a**) Change of the contact angle and (**b**) the drop base radius with time for the Mg/TiC system [87, 88]

Laurent et al. [89] pointed out that the most probable mechanism for the disappearance of the aluminum oxide film could be the dissociation of the aluminum oxide by liquid aluminum, involving the formation of a gaseous suboxide, Al_2O, which depends on the holding time and temperature, according to the reaction:

$$4Al_{(l)} + Al_2O_{3(s)} = 3Al_2O_{(g)} \tag{2.18}$$

For such a reaction to take place, it must be assumed that the oxide film is very thin and that the oxide layer is not compact upon the initial melting of the aluminum. Brennan and Pask [90] suggested that the oxide layer disappears completely at about 870 °C.

Figure 2.16 shows the variation of the contact angle θ and the drop base radius for pure Mg on TiC substrates with the time in a different temperature. A non-wetting

Fig. 2.17 Wetting behavior of Al/TiC and Mg/TiC in function of the temperature after 30 min [87, 88]

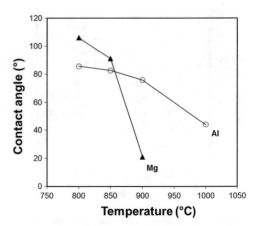

behavior is evident at 800 and 850 °C, but increasing the temperature to 900 °C spontaneous wetting is obtained, reaching angles of about 10° after 35 min.

The shorter time of the tests was due to the high evaporation of Mg. In the cases of AlMg-alloys, the chemical composition can change with the time due to the evaporation of Mg, even though using Ar minimizes the evaporation of Mg. The excessive evaporation of Mg was observed even at 800 °C under static argon and atmospheric pressure.

A decrease of θ is observed, while the spreading base radius R increases and the height of the drop decreases. In the case of the Mg/TiC at 900 °C, the plot of the contact angle with time showed two main stages, one where a rapid decrease of θ with time occurred and the other corresponding to slow rate of decreases of θ, but not reaching a "steady state." This case is a nonreactive system, and the spreading rate is controlled by the viscous flow, reaching a maximum of R, after which there is a decrease due to the high evaporation of Mg. These systems are characterized by extremely rapid kinetics of wetting and very weak temperature dependence variations of contact angle changing from a non-wetting behavior, to wetting behavior between 850 and 900 °C as is shown in Fig. 2.17. According to the present results, the transition of non-wetting to wetting behavior for the Al/TiC system is below 800 °C.

2.10.3 Interfacial Characterization

After the Al and Mg drop solidified on solid TiC substrate (Fig. 2.18), they were cross-sectioned in order to analyze the interface and perform the characterization. In reactive systems (Al/TiC), wetting of the liquid metal on the solid substrate frequently occurs with extensive chemical reaction and the formation of a new solid compound at the metal/substrate interface.

Fig. 2.18 Aluminum drop
after been solidified on TiC
substrate

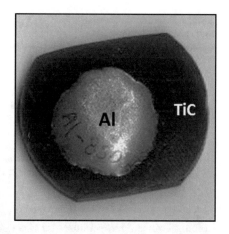

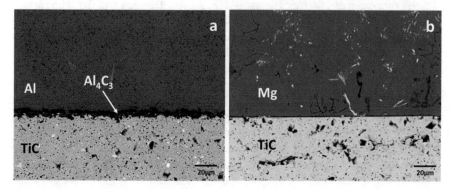

Fig. 2.19 Cross sections of the sessile drop-substrate interface, (**a**) Al/TiC and (**b**) Mg/TiC at
900 °C, after 120 and 35 min, respectively

Just below the aluminum drop, a region of major porosity in the substrate was
observed; this could be possible due to the dissolution of TiC by Al and diffusion of
Ti through the interface. The interfacial reaction produces a strong decrease in
contact angle from about 120° to 51°.

Figure 2.19 shows cross sections of the drop-substrate interfaces for Al/TiC and
Mg/TiC assemblies. In reactive systems, wetting frequently occurs with extensive
chemical reaction and the formation of a new solid compound at the metal/
substrate interface. For Al/TiC assembly it is clearly observed the formation of a
new compound in the interface (Al_4C_3) with around 5–8 μm of thickness
(Fig. 2.19a), however because of the reduced volume of the interface, it was not
possible to examine accurately cross-sectional samples of the metal-ceramic couples by XRD. However, quantitative determinations performed by electron probe
microanalysis EPMA along different positions of the interface revealed the presence of elementary Al and C with chemical stoichiometries suggesting an Al_4C_3
compound.

Table 2.3 Work of adhesion evaluated in function of the surface tension and the contact angle

Temperature (°C)	Mg/TiC (mJ/m²)[a]	Al/TiC (mJ/m²)[b]
800	435	1317
850	527	1349
900	904	1405
1000	–	1608

[a]The contact angle used was the value obtained after 45 min, when the contact angle is nearly constant, except at 900 °C that corresponds to 15 min.
[b]The contact angle used was the value obtained after 120 min, when the contact angle is nearly constant, except at 1000 °C that corresponds to 50 min.

The thickness of the reaction layer varied within the samples depending of the holding time and temperature and was not uniform in size. In Mg/TiC system, there was not formation of interfacial reaction (Fig. 2.19b). The nature of the solid surface is not significantly modified by its contact with the metallic phase. Thus by wetting experiments, it was confirmed that magnesium does not form stable carbides, so carbides like TiC should be suitable reinforcements for Mg matrix composites.

The work of adhesion of the metal-ceramic couples was calculated in function of the contact angle (θ) and surface tension (γ_{LV}), using Eq. (2.4). The values of γ_{LV} were calculated using data in reference [91]. Since the work of adhesion is the work per unit area of interface that must be performed to separate reversibly the two phases, it is therefore a measure of the strength of the binding between the phases [9, 14, 92]. Table 2.3 shows the results obtained for both systems.

The condition for perfect wetting is given by $W_a \geq 2\gamma_{LV}$. This means that the adhesion energy between the ceramic and the melt should be more than twice the surface tension of the liquid. The W_a for Mg/TiC system calculated at 800 °C was 435 mJ/m². Considering that $\gamma_{LV(Mg)} = 538$ mJ/m², this means about 40% of the cohesion work of liquid Mg, which suggests that Mg/TiC interface will be weak energetically. In this nonreactive system, the adhesion could be attributed to van der Waals forces mainly.

In the case of Al/TiC at the same temperature, the work of adhesion represents 78% of the cohesion work of liquid Al, which indicate a strong interface.

As the mechanical behavior of solidified drops under thermal stresses produced during cooling can be qualitatively characterized by observing the type of fracture, the solidified metal drops were microscopically examined. In the Mg/TiC systems, adhesive fracture as is shown in Fig. 2.20a was observed (interfacial rupture path corresponding to a weak interface), while Al/TiC system showed cohesive fracture in most of the cases (bulk rupture path in TiC near the interface) corresponding to a strong interface as is shown in Fig. 2.20b. Thus, adhesion in Al/TiC cannot be interpreted in terms of weak interactions of van der Waals but can be interpreted in terms of strong chemical interactions, such as covalent bond.

Since reaction systems are characterized by pronounced changes of θ with time and strong temperature-dependent variations, it was possible to calculate the activation energy for the wetting of Al/TiC by performing a kinetic study from the

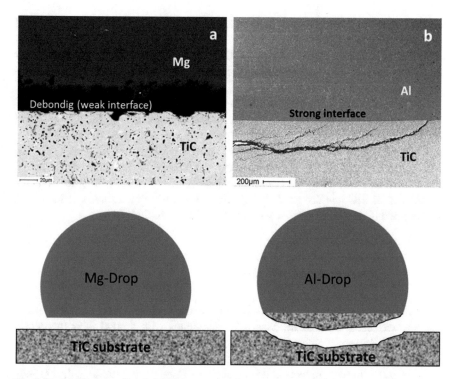

Fig. 2.20 (a) Weak bonding in TiC/Mg systems, (b) strong joining in TiC/Al systems

spreading of the drop base radius versus time (dr/dt) applying the Arrhenius equation [52, 93, 94]. Generally speaking, high values of activation energy are associated with chemical interaction [52, 94, 95]. In the present work, the apparent activation energy obtained for Al/TiC was 117 kJ/mol, confirming that wetting is driven by reaction between Al and TiC. Meanwhile the activation energy for the TiC/Mg system was 360 kJ/mol.

2.11 Wettability of TiC by Pure Cu

The wetting behavior between molten pure Cu and solid TiC substrates was studied by the sessile drop technique in the temperature range of 1100 to 1130 °C [96]. For this purpose, the contact angles of liquid Cu on sintered TiC substrates were measured by the sessile drop technique under a static atmosphere of argon.

TiC/Cu composites are extensively employed in applications where a good electrical and/or thermal conductivity is required [97, 98]. However some properties like wettability must be evaluated in order to improve the fabrication and performance of the composites. As copper has very high melting point (1083 °C), the use of Cu as matrix in metal matrix composites has been less, as well as the studies of wetting are more difficult.

One of the most complete studies in TiC/Cu composites was performed by Akhtar et al. [99]. They study the processing, microstructure, mechanical properties, electrical conductivity, and wear behavior of TiC/Cu composites.

The first studies of wetting of TiC by Cu were carried out by Rhee on 1970 [73]. Contact angles between 101 and 126° were obtained for the TiC/Cu system at temperatures of 1100–1180 °C. Over the past years, wetting behavior of TiC by Cu and several alloys has been the subject of many investigations [56, 81, 100]. The wetting behavior in the TiC/(Cu-Al) system was studied by the sessile drop method over the entire Cu-Al alloy concentration. However, few studies with pure Cu on the wettability of TiC has been carried out [56, 73, 81, 88, 96]. In most of the cases, wetting of TiC has been studied with copper alloys [88, 96, 101].

2.11.1 Experimental Conditions

Titanium carbide powder (H. C. Starck, grade c.a.s.) with an average particle size of 1.2 μm was used to fabricate dense TiC substrates with around 96% theoretical density and 25 mm diameter and smooth surfaces with a 2.6–2.8 nm roughness. Electrolytic grade Cu was used in small cubic shape (~0.8 g); the cube of the alloy was cleaned with sodium hydroxide and acetone. Previous to introduce the argon to the chamber, a high vacuum was obtained (1×10^{-5} Torr) to remove the maximum the oxygen.

The wetting experiments were carried out under ultrahigh purity argon (99.99%) at atmospheric pressure at 1000 and 1130 °C in an experimental setup specially designed for this purpose as was shown in Fig. 2.13. The change in contact angle and drop base radius was recorded photographically at various time intervals until 180 min. Figure 2.21 shows several images of the sessile drop system for TiC/Cu assemblies at 1100 and 1130 °C during 180 min.

2.11.2 Wetting and Spreading Kinetic

Experiments with pure copper at 1100 and 1130 °C were carried out. The results of the contact angle and drop base radius versus time are shown in Fig. 2.22. A non-wetting behavior after long periods of time (180 min) was observed for this system. The contact angle for both temperatures used was above 105° after 180 min. Few studies concerning wetting of TiC by Cu exists, and very much work has been done on wetting with Al-Cu alloys. The first studies about TiC/Cu wetting were obtained in the early 1970s by Rhee [73]. They obtained contact angles around 126°, 115°, and 105° for 1100, 1130, and 1170 °C. Later in 1973 Mortimer and Nicholas [101] studied the wetting of carbon and some carbides by copper alloys. More recently, Zarrinfar et al. [102, 103] reported poor wetting between molten copper

Fig. 2.21 Change in contact angle recorded photographically at various time intervals until 180 min for the TiC/Cu system

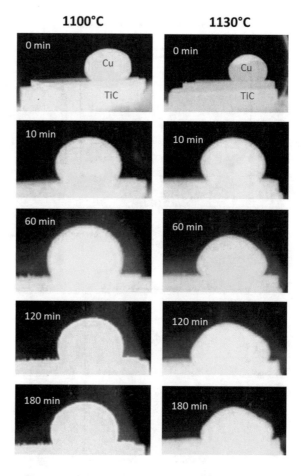

and TiC which makes it difficult to produce copper-based ceramics composites. However, a previous work reported the fabrication of AlCu-alloys/TiC composites processed by the pressureless infiltration technique at temperatures of 900 and 1000 °C with copper content until 33% [104]. The poor wettability is attributed to the stable electron configuration of copper which has a full "3d" orbital. It has been suggested that copper behaves as a donor rather than an acceptor of electrons. Froumin et al. [100] reported that oxidation treatment improves the wettability of TiC by Cu-Al alloys. They suggested that wetting by pure Cu is controlled by dissolution of the TiC phase, leads to enhance transfer of Ti into the melt, and thereby improves wetting.

Mortimer and Nicholas [101] studied the wetting of TiC by Cu in a range of hypo-stoichiometry (Fig. 2.23a). Meanwhile, Xiao and Derby [81] studied the wetting behavior of Ag and Cu on TiC in a wide range of stoichiometries. They found that Ag and Cu wet TiC at hypo-stoichiometry compositions of $TiC_{0.6}$ as is observed in Fig. 2.23b. In this way, it is demonstrated the hypothesis that it is the metallic character of a ceramic that promotes wetting by metals.

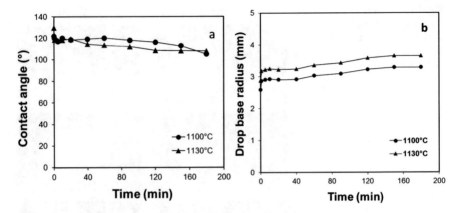

Fig. 2.22 (**a**) Kinetic spreading of the contact angle and (**b**) drop base radius in function of the time for TiC/Cu system [96]

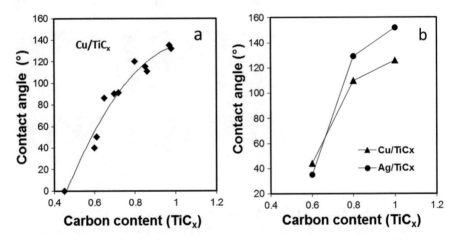

Fig. 2.23 (**a**) Contact angle for Cu on TiC as a function of stoichiometry [101]. (**b**) Contact angle of liquid Cu at 1150 °C and Ag at 1050 °C on TiC with different stoichiometry [81]

2.11.3 Interfacial Characterization

The solidified drops were sectioned and polished to examine the possible interfacial reactions using scanning electron microscopy (SEM). For the pure-Cu/TiC system, no reaction was observed as is shown in Fig. 2.24. Similar results were reported by several researchers [56, 73]. Thus wetting of TiC by Cu seems to occur only in the presence of some interfacial reaction or adding Ti to the molten metal. The additions of Ti in the molten copper improve wetting of TiC substrates.

As a measure of the joining between the ceramic and the metal, the work of adhesion was calculated in function of the contact angle (θ) and surface tension

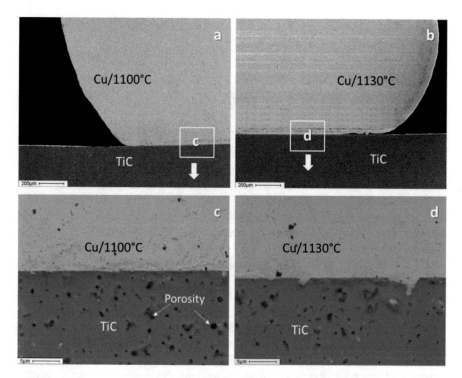

Fig. 2.24 Cross section of the drop-substrate interface for the TiC/Cu system obtained at (**a**) 1100 °C and (**b**) 1130 °C; (**c**) and (**d**) high magnification of selected area on the image (**a**) and (**b**) respectively

Table 2.4 Adhesion and cohesion work evaluated in function of the surface tension (γ_{LV}) and the contact angle (θ) obtained after reach the equilibrium (180 min)

System	Temperature (°C)	Adhesion work (mJ/m²)			Cohesion work (%)
		γ_{LV}	θ	W_a	
Pure-Cu/TiC	1100	1287	116	723	28
	1130	1282	108	886	34

(γ_{LV}), using the Eq. (2.4), where the values of γ_{LV} were calculated using data in reference [91]. Table 2.4 shows the results obtained. The work of adhesion is a crucial parameter in processing of MMCs and on the interfacial properties of the materials which is related to the wetting. Since the work of adhesion is the work per unit area of interface that must be performed to separate reversibly the two phases, it is therefore a measure of the strength of the binding between the phases.

The condition for perfect wetting is given by $W_a \geq 2\gamma_{LV}$. This means that the adhesion energy between the ceramic and the melt should be more than twice the surface tension of the liquid. The W_a for pure Cu/TiC system calculated at 1100 °C was 723 mJ/m². Considering that $\gamma_{LV(Cu)} = 1287$ mJ/m², this means about 28% of the cohesion work of liquid Cu (Table 2.4), which suggests that Cu/TiC interface will be weak energetically. In this nonreactive system, the adhesion could be attributed to van der Waals forces mainly.

2.12   Wettability of TiC by Al-Cu Alloys

The wetting behavior and the interfacial reactions that occurred between molten Al-Cu alloys (1, 4, 8, 20, 33, and 100 wt.% Cu) and solid TiC substrates were studied by the sessile drop technique at 800, 900, and 1000 °C [96]. The effect of wetting behavior on the interfacial reaction layer was studied. For this purpose, the contact angles of liquid Al-Cu alloys on sintered TiC substrates were measured by the sessile drop technique under a static atmosphere of argon, and the relationship between the nature of the reaction products and the contact angle was investigated. The spreading kinetics and the work of adhesion were evaluated. This investigation shows the findings of a study undertaken to determine the effects of time and chemical composition on the interfacial reactions. The findings led to a new insight of the wetting process of TiC by Al-Cu alloys which are applicable to any liquid metal-ceramic reactive system. The results have important implications for the liquid-state processing technology of Al-Cu alloys/TiC composites.

Wetting behavior of TiC by Cu and several alloys has been the subject of many investigations [56, 73, 81, 100, 101]. The wetting behavior in the TiC/(Cu-Al) system was studied by the sessile drop method over the entire Cu-Al alloy concentration [100]. Similar studies were performed by Frage et al. [56] in the wetting of TiC/Cu system, suggesting that partial dissolution of Ti into the molten metal improves wetting permitting to reach an equilibrium contact angle close to 90°. The wetting of TiC by Cu and Ti-Cu alloys with a range of stoichiometry has been studied using the sessile drop method [100, 101]. They found that Cu wets TiC below hypo-stoichiometry compositions of $TiC_{0.6}$. More recent studies on spreading kinetics revealed that a given system is strongly affected by the experimental conditions. The same system exhibits different kinetic patterns under different conditions [93].

2.12.1   Experimental Conditions

Titanium carbide powder (H. C. Starck, grade c.a.s.) with an average particle size of 1.2 μm was used to fabricate dense TiC substrates with around 96% theoretical density and 25 mm diameter and smooth surfaces with a 2.6–2.8 nm roughness. Hot-pressing was carried out at the temperature of 1800 °C using a constant pressure of 30 MPa for 30 min. Pure Al (99.99%) and electrolytic grade Cu were used to fabricate the Al-Cu alloys. Pieces were cut in cubic shape (~0.6 g); the cube of the alloy was cleaned with sodium hydroxide and acetone. Previous to introduce the argon to the chamber, a high vacuum was obtained (1×10^{-5} Torr) to remove the maximum the oxygen.

The wetting experiments were carried out using Al-Cu alloys containing 1, 4, 8, 20, and 33 wt.% of Cu, under ultrahigh purity argon (99.99%) at atmospheric pressure between 800 and 1000 °C in an experimental setup specially designed for this purpose as was shown in Fig. 2.13. The change in contact angle and drop base radius was recorded photographically at various time intervals until a steady state was reached. The solidified drops were sectioned and polished to examine the possible interfacial reactions using scanning electron microscopy (SEM), energy-dispersive analysis (EDS), and electron probe microanalysis (EPMA).

2.12.2 Wetting and Spreading Kinetic

The change in contact angle and drop base radius was recorded photographically at various time intervals until 150 min. Figure 2.25 shows several images of the sessile drop system for TiC/Al-Cu alloy assemblies at 800, 900, and 1000 °C. The main changes in the drop shape and contact angle take place within the first 15–30 min as can be observed in Fig. 2.25.

The time dependence of the contact angle and the drop base radius measured in the 800–1000 °C temperature range for the AlCu-alloys on TiC substrates are shown in Fig. 2.26. The main changes in the drop shape and contact angle take place within the first 15–30 min as can be observed in Fig. 2.25.

These curves exhibit different behavior depending on the temperature and copper content, and present three characteristic regions of the contact angle with respect to time: the first typified by a sharp slope (deoxidation of the Al drop) where the contact angle decreased at a specific rate, the second presenting a slighter slope (chemical reaction), and the third where the value of θ is nearly constant. This behavior is more noticeable at 900 and 1000 °C. More than 120 min were generally required to obtain a nearly constant angle (between 50 and 60°) depending on the temperature. Good wettability in all the cases was observed, increasing the spreading radius with time in response to contact angle decrease, until reaching a nearly constant radius value. The non-wetting to wetting transition of a low temperature (800 °C) requires more than 60 min, and in the case of 4 wt.% Cu requires more than 90 min. From Fig. 2.26 it is clearly observed a non-wetting to wetting transition below 800 °C. However, this transition is time dependent. Good wetting was obtained at 800 °C after 60 min for all copper contents, except for Al-4Cu where needed more than 90 min to have good wetting ($\theta < 90°$).

The effect of Cu in the change of the contact angle at isothermal conditions is shown in Fig. 2.27. It was observed that Al-Cu alloys exhibited different behavior depending of the temperature and the copper content. At lower temperatures (800–900 °C), the wetting increases increasing the copper content, except for Al-1Cu where the wetting behavior in all the cases was similar to the pure-Al. This behavior can be attributed that by increasing the copper content in the aluminum, the melting point of the alloy decreases considerably (from 660 for pure Al to 548 °C in the eutectic composition, 33 wt.% Cu). This gradient of temperature ($\Delta T \sim 112$ °C) has more noticeable effect than the increases in surface tension [96].

At higher temperature (1000 °C) the wetting decreases slightly increasing the copper content, which can be related to the increases in the viscosity and the surface tension of the alloy [105], and the gradient of temperature has minor effect.

Because of the rapid change in the contact angle during the first moments of wetting, more accurate determination of the initial contact angle is possible using a logarithmic scale for time. Figure 2.28 shows the contact angle curves plotted on a logarithmic time scale. When the contact angle curves were plotted on a logarithmic time scale, at the temperature of 800 and 900 °C, the contact angles can be seen to

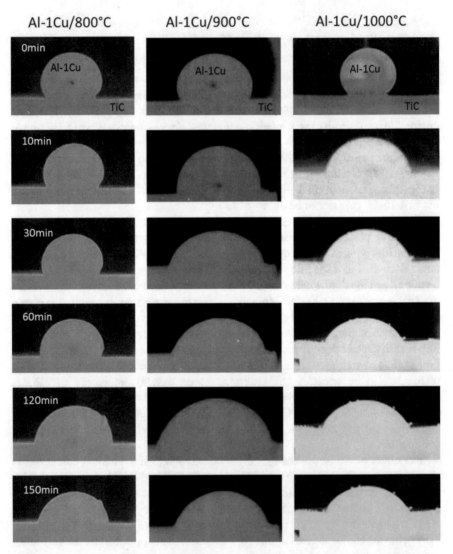

Fig. 2.25 Change in contact angle recorded photographically at various time intervals until 150 min for Al-1Cu/TiC and Al-20Cu/TiC

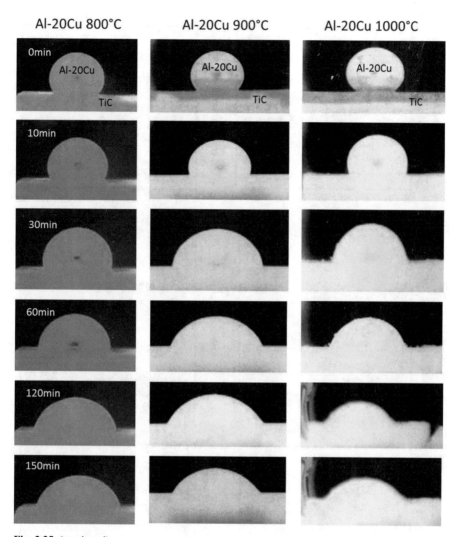

Fig. 2.25 (continued)

progress through the two phases, (I) deoxidation of the Al drop and (II) interfacial reaction wetting phase, where the second phase does not reach the equilibrium; this could be related to the stationary regime that has not yet been reached at this temperature and would need longer period of time to reach the equilibrium. At 1000 °C the contact angle seems to decrease through three phases: (I) deoxidation of the Al drop (this period is shorter than at 800 and 900 °C and depends of the chemical composition), (II) wetting through interfacial reaction, and (III) nearly constant angle.

According to Fig. 2.28, the period of the phase I depends of the temperature and the chemical composition of the alloy used. The Al-Cu alloys/TiC at 800 °C

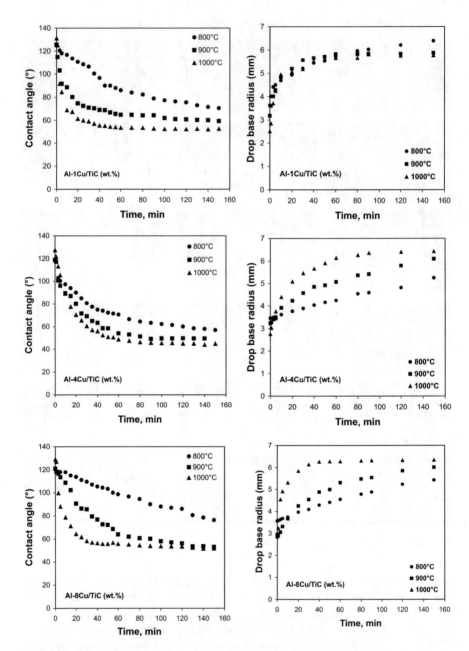

Fig. 2.26 Kinetic spreading of the contact angle and the drop base radius with the time for the Al-Cu alloys/TiC systems [96]

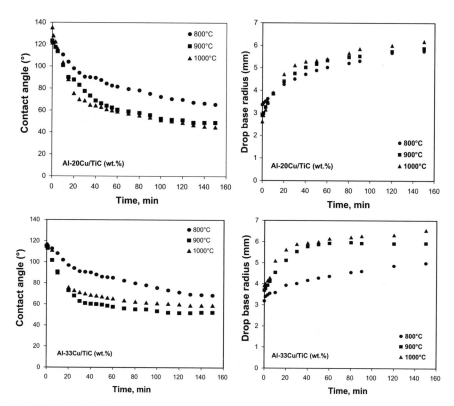

Fig. 2.26 (continued)

presented the longer phase I. Therefore, the change from the phase I to phase II correspond to a particular development in the interface reaction.

The results obtained from sessile drop experiments indicated that spreading rate increases with the time and depends of the alloy and temperature used. The wettability of ceramic surfaces by metals and alloys is governed by several chemical and physical processes. The degree of reaction will depend of the temperature and chemical composition of the alloy used. The value of the contact angle can be significantly reduced by certain alloying elements which form continuous layers of compounds at the interface, by reaction with the solid substrate [49–55, 75, 106].

The wettability of ceramics surfaces by metallic melts is strongly dependent on the stoichiometry of the ceramic compound. TiC has a wide range of stability, and its chemical, physical, and mechanical properties largely depend on the stoichiometry [75, 81, 102].

The reactive systems are characterized by pronounced changes of θ with time and strong temperature-dependent variations; these results make possible to calculate the activation energy for the wetting of Al-Cu alloys/TiC by performing a kinetic study from the spreading of the drop base radius versus time (dr/dt) applying the Arrhenius

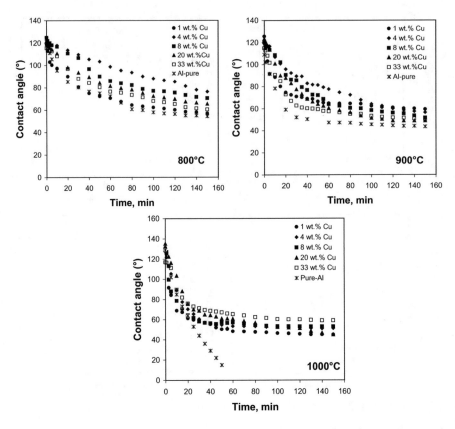

Fig. 2.27 Effect of Cu content in the contact angle for the Al-Cu alloys at isothermal conditions [96]

equation as is shown in Fig. 2.29. The high values of activation energy indicates that spreading is not a simple viscosity controlled phenomenon but is a chemical reaction process [94, 50–55]. In the present work, the average apparent activation energy obtained for Al-Cu alloys/TiC systems was 86 kJ/mol, confirming that wetting is driven by chemical reaction at the interface. Table 2.5 shows the activation energy values obtained from the Arrhenius plot of Fig. 2.29.

2.12.3   Interfacial Characterization

In reactive systems, good wetting occurs simultaneously with the formation of a new solid phase at the metal/substrate interface. In nonreactive systems [87, 106], the spreading rate is controlled by the viscous flow and described (for $\theta < 60°$) by a power function of drop base radius versus time [49].

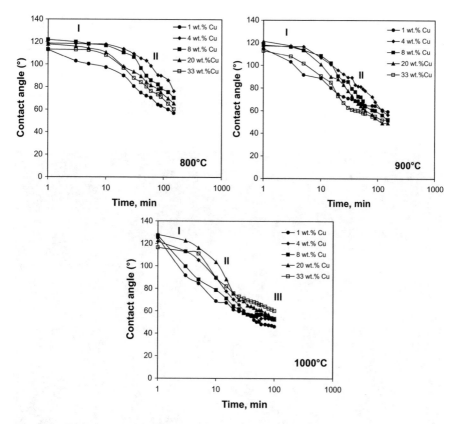

Fig. 2.28 Contact angle curves in a logarithmic time scale [96]

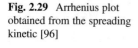

Fig. 2.29 Arrhenius plot obtained from the spreading kinetic [96]

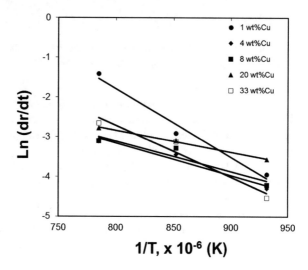

Table 2.5 Activation
energies

Alloy (wt.%)	E_a (kJ/mol)
Al-1Cu/TiC	145
Al-4Cu/TiC	68
Al-8Cu/TiC	63
Al-20Cu/TiC	45
Al-33Cu/TiC	108

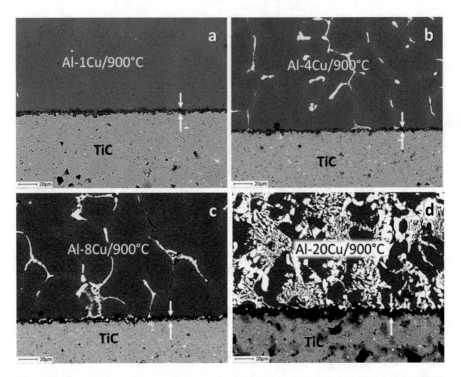

Fig. 2.30 Cross sections of the sessile drop-substrate interface obtained at 900 °C. (**a**) Al-1Cu/TiC, (**b**) Al-4Cu/TiC, (**c**) Al-8Cu/TiC, and (**d**) Al-20Cu/TiC [96]

Figure 2.30 shows cross sections of the drop-substrate interfaces for Al-Cu alloys/ TiC. The presence of a new reaction product is clearly observed for all the systems. The thickness of the reaction layer varied with the samples and was discontinuous in nature, particularly for the high Cu contents.

Figure 2.31 shows the thickness of the reaction layer in function of the Cu content. The reactivity at the interface seems to be greater at high copper contents (20 wt.%). At higher copper contents, it is suggested that presence of copper promotes the diffusion of Ti and C to the interface leaving a considerably quantity of porosity under the drop as is observed in Fig. 2.30d. Some studies performed by several researchers reported that Ti not only improves wetting by increasing the

Fig. 2.31 Effect of Cu content in the thickness of the interfacial layer at 900 °C [96]

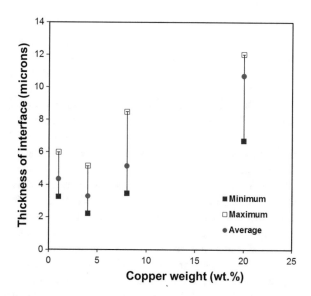

Table 2.6 Elemental analysis performed in the interface in samples tested at 900 °C

Alloy (wt.%)	Elements detected at the interface	Possible reaction products
Al-1 Cu	Al, C, Cu, O	Al_4C_3, Al_2O_3, $CuAl_2$, $TiAl_3$
Al-4 Cu	Al, C, Cu, O	Al_4C_3, Al_2O_3, $CuAl_2$, $TiAl_3$, Ti_3AlC
Al-8 Cu	Al, C, Cu, O	Al_4C_3, Al_2O_3, $CuAl_2$, $CuAl_2O_4$, $TiCu_3TiCu_4$
Al-20Cu	Al, C, Cu, O, Ti	Al_4C_3, $CuAl_2O_4$ Ti_3Al, $TiAl_3$, $CuAl_2$, $TiCu_3TiCu_4$, $TiCu$
Al-33Cu	Al, C, Cu, O, Ti	Al_4C_3, $CuAl_2O_4$ Ti_3Al, $TiAl_3$, $CuAl_2$, $TiCu_3TiCu_4$, $TiCu$

tendency for Cu to accept electrons but also because Ti has been transferred to the melt [102, 103]. Traces of Ti were identified in the interface near the aluminum, which suggested the dissolution or dissociation of TiC and Ti diffuse to the interface; this was corroborated by the presence of some titanium aluminides and some other compounds with Cu as is shown in Table 2.6.

Quantitative determinations performed by electron probe microanalysis (EPMA) along different regions of the interface revealed the presence of elementary Al and C with chemical stoichiometries suggesting an Al_4C_3 compound. Long periods of time on wetting permit the reaction of the reinforcement and the molten metal. The formation of Al_4C_3 at the interface improves the wetting but has several undesirable effects. Al_4C_3 dissolves in water, degrading the corrosion behavior of the composite [107].

The spreading of the aluminum drop is observed to occur according to the formation of Al_4C_3, $CuAl_2O_4$, $CuAl_2$, and $TiCu_x$, mainly leading to a decrease in the contact angle.

The ceramic surface initially flat and smooth appears rougher after performed the test, because of the dissociation of TiC leading to the formation of Al_4C_3. In a region about 100–1000 μm (depending of the sample and temperature of the test) just below to the drop, the diffusion effect can be observed in the roughness of the substrate leaving a big quantity of microvoids as can be observed in Fig. 2.30. Far of this region, the substrate continues to being flat without the presence of microvoids. Thus, it is believed that formation of Al_4C_3 at the interface is greatly affected by the diffusion of carbon when TiC is dissociated; this is influenced by the temperature and time of the test. This diffusion obviously will be greater at higher temperatures, and the presence of high contents of copper enhances the diffusion. Thus, it is observed a rougher and high quantity of microvoids with high copper contents (Fig. 2.30d). The time of the tests has effect on the formation of these compounds and carbide and also during the cooling, the reaction continues in a solid state. In addition, the precipitation of intermetallic phases delays the formation of Al_4C_3. In all the systems, the presence Al_4C_3 in the interface contributes to having better wetting. However, some authors affirm that in this case, the wetting is on the product of the reaction layer and not in the ceramic substrate [107–109].

The mechanical behavior of the solidified drops under thermal stresses produced during cooling was characterized by observing the type of fracture of some Al/TiC assemblies. In some cases, especially at high Cu contents, a cohesive fracture (bulk rupture path in TiC near the interface) was observed corresponding to a strong interface as is shown represented schematically in Fig. 2.32. The difference in the thermal expansion coefficient between the metal and ceramic substrate is responsible of this type of fracture.

As a measure of the joining between the ceramic and the metal, the work of adhesion was calculated in function of the contact angle (θ) and surface tension (γ_{LV}), using the Eq. (2.4), where the values of γ_{LV} were calculated using data in reference [91]. Table 2.7 shows the results obtained. The work of adhesion is a crucial parameter in processing of MMCs and on the interfacial properties of the materials which is related to the wetting. The contact angle used in the calculation of adhesion work was the value obtained after 120 min, because at this time the contact angle is nearly constant.

Since the work of adhesion is the work per unit area of interface that must be performed to separate reversibly the two phases, it is therefore a measure of the strength of the binding between the phases.

Fig. 2.32 Typical schematic cohesive fracture observed at high Cu contents

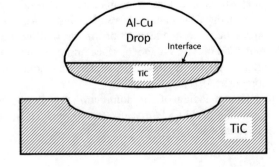

Table 2.7 Work of adhesion evaluated in function of the surface tension (γ_{LV}) and the contact angle (θ) obtained after reach the equilibrium (120 min)

| | Adhesion work (mJ/m²) | | | | | | | | |
| | 800 °C | | | 900 °C | | | 1000 °C | | |
System	γ_{LV}	θ	W_a	γ_{LV}	θ	W_a	γ_{LV}	θ	W_a
Al-1Cu/TiC	876	60	1314	860	61	1277	844	53	1352
Al-4Cu/TiC	956	86	1023	938	60	1407	920	45	1571
Al-8Cu/TiC	962	76	1195	944	56	1472	926	53	1483
Al-20Cu/TiC	977	70	1311	963	49	1595	949	51	1546
Al-33Cu/TiC	997	61	1480	930	52	1503	919	65	1307

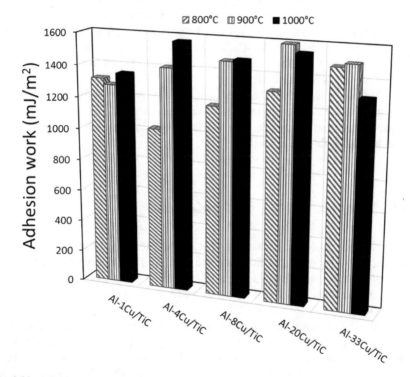

Fig. 2.33 Adhesion work in function copper content and temperature

Wettability and work of adhesion between liquid metals and nonmetallic solids can be increased by inducing an extensive chemical reaction at the interface. Figure 2.33 shows the behavior of the work of adhesion for the system AlCu$_x$-/TiC system in function of the temperature. It is clearly observed that work of adhesion has a trend to increase as the temperature increases, consequently decreasing the contact angle with the time.

Table 2.8 Percentage of the cohesion work at the interface in function of the surface tension

System	Cohesion work (%)		
	800 °C	900 °C	1000 °C
Al-1Cu/TiC	75	74	80
Al-4Cu/TiC	53	75	85
Al-8Cu/TiC	63	78	80
Al-20Cu/TiC	67	83	81
Al-33Cu/TiC	74	81	74

The condition for perfect wetting is given by $W_a \geq 2\gamma_{LV}$. This means that the adhesion energy between the ceramic and the melt should be more than twice the surface tension of the liquid. The W_a for Al-1Cu/TiC system calculated at 1000 °C was 1352 mJ/m^2. Considering that $\gamma_{LV(Cu)} = 844$ mJ/m^2, this means the work of adhesion represents 80% of the cohesion work of liquid alloy indicating a strong interface. Thus, adhesion in Al-1Cu/TiC system cannot be interpreted in terms of weak interactions of van der Waals but can be interpreted in terms of strong chemical interactions, such as covalent bond.

It is obvious that good wetting does not necessarily mean a thermodynamically strong interface [110], as can be seen of the results obtained in Table 2.8. Good wetting means that the interfacial bond is energetically nearly as strong as the cohesion bond of the liquid itself [111].

2.13   Wettability of TiC by Al-Mg Alloys

Wettability and interfacial reactions that occur between molten Al-Mg alloys containing 1, 4, 8, and 20 wt.% of Mg and solid TiC substrates were studied. The sessile drop technique was used to study the wetting behavior of Al-Mg alloys on TiC sintered ceramic substrates under argon in the temperature of 750, 800, and 900 °C [112].

The wetting of ceramics by molten metals usually involves interfacial reactions [113, 114]. The interface reaction affects the composition of the liquid, the volume of the liquid drop, and the size of the reaction zone. A change in any of these parameters will result in a change in the wetting process. Since the interface reaction is dependent not only on temperature but also on time, the wetting process changes as time passes even in isothermal conditions.

The interfacial reactions produce new compounds at the interface and change the composition of the liquid metals and ceramics. As a result, all of the solid-liquid, liquid-vapor, and solid-vapor interfacial free energies change. In a reactive system, interfacial reactions often change the degree of wetting over time and have led many researchers to misinterpret some aspects of wetting [115].

One of the main functions of Mg is to reduce the surface tension and the viscosity of aluminum alloys [21]. Titanium carbide as reinforcement is thermodynamically stable in contact with pure molten magnesium [87]. However, if Mg is alloyed with elements such as Al, a reaction to form MgO or $MgAl_2O_4$ will proceed [107, 108].

The extent of the chemical reaction and the type of reaction products formed are dependent on the processing temperature, pressure, atmosphere, matrix composition, and surface chemistry of reinforcements.

This investigation shows the findings of a study undertaken to determine the effects of time and chemical composition on the interfacial reactions. For this purpose, the contact angles of liquid AlMg-alloys on sintered TiC substrates were measured by the sessile drop technique and the relationship between the nature of the reaction products and the contact angle was investigated. In addition, the effect of Mg loss from the system on the kinetics of the wetting phenomena was investigated. The findings led to a new insight of the wetting process of TiC by Al-Mg alloys which are applicable to any liquid metal-ceramic system where chemical reaction occurs.

2.13.1 Experimental Conditions

Titanium carbide powder (H. C. Starck, grade c.a.s.) with an average particle size of 1.2 μm was used to fabricate dense TiC substrates with 96% theoretical density. The 25-mm-diameter TiC substrates were prepared by hot-pressing under vacuum in a graphite furnace (Fig. 2.12). The temperature was 1800 °C using a constant pressure of 30 MPa for 30 min. The substrates were polished on one face using 1 μm diamond paste for the final polish and cleaned ultrasonically in acetone. Results obtained from AFM indicate that the grinding-polishing route followed on TiC substrates yielded smooth surfaces with a 2.6–2.8 nm roughness range. Pure Al (99.99%) and electrolytic-grade Mg were used to fabricate the Al-Mg alloys (1, 4, 8, and 20 wt. % of Mg). Pieces were cut in cubic shape (~0.6 g), and surface is finished in the same way than the substrates in preparation for wetting tests.

The wetting experiments were conducted using Al-Mg alloys containing 1, 4, 8, and 20 wt.% of Mg, under ultrahigh purity argon (99.99%) at atmospheric pressure and 750, 800, and 900 °C in an experimental setup which was shown in Fig. 2.13. The change in contact angle and drop base radius was recorded photographically at various time intervals until a steady state was reached. After finishing the experiment, the samples were immediately removed from the hot zone inducing fast cooling. The solidified drops were sectioned and polished to examine the interfacial phases using scanning electron microscopy (SEM), energy-dispersive analysis (EDS), and electron probe microanalysis (EPMA).

2.13.2 Wetting and Spreading Kinetic

The drop shape to measure the change in contact angle and drop base radius was recorded photographically at various time intervals until 120 min. Figure 2.34 shows several images of the sessile drop for Al-1 Mg/TiC and Al-20Mg/TiC assemblies at 750, 800, and 900 °C.

Figure 2.35 shows the variation of the contact angle and the drop base radius for the Al-Mg alloys on TiC substrates as a function of time and temperature. These curves exhibit different behavior depending on the temperature and Mg content, and present three characteristic regions of the contact angle with respect to time: the first typified by a sharp slope (deoxidation of the Al drop) where the contact angle decreased at a specific rate, the second presenting a slighter slope (chemical reaction), and the third where the value of θ is nearly constant. This behavior is more noticeable at 900 °C. More than 120 min were generally required to obtain a nearly constant angle. Good wettability was observed, increasing the spreading radius with

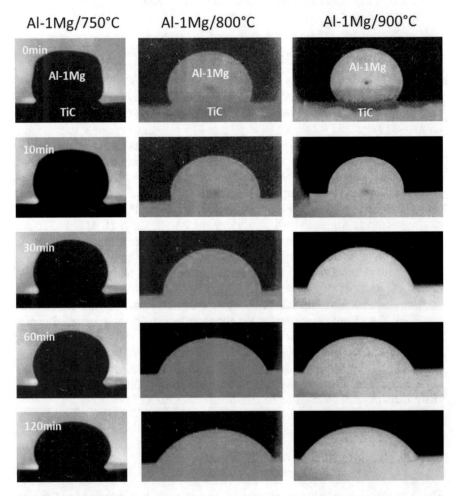

Fig. 2.34 Change in contact angle recorded photographically at various time intervals until 120 min for Al-1Mg/TiC and Al-20Mg/TiC

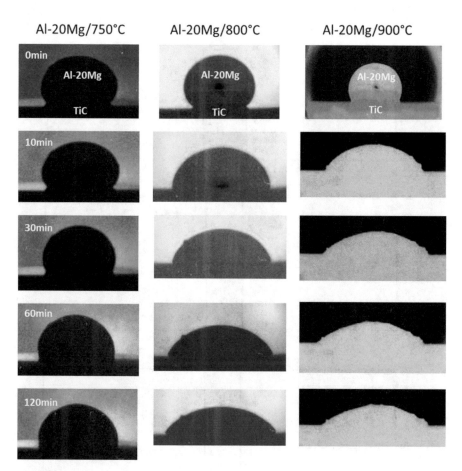

Fig. 2.34 (continued)

time in response to contact angle decrease. However, at lower temperatures (750 °C) for 1 and 4 wt.% of Mg, a non-wetting behavior was observed even after long periods of time (150 min).

The effect of Mg in the change of the contact angle at isothermal conditions is shown in Fig. 2.36. It was observed that by increasing the Mg content, the contact angle decreases. The profile of the contact angle for pure Al was included as reference to observe the effect of Mg in the aluminum. By increasing the Mg content, the surface tension decreases significantly; therefore the wettability increases (Al-1wt.% Mg). However, by increasing the Mg until 4 and 8 wt.%, the oxidation effect is greater than decreases in surface tension, obtaining worse wetting, but by increasing the Mg until 20 wt.%, effect of surface tension is bigger, obtaining the best wetting with this alloy. The above surface tension reduction is very sharp for the initial 1 wt.% Mg addition; however, with further increase in Mg content, the reduction is very marginal [22].

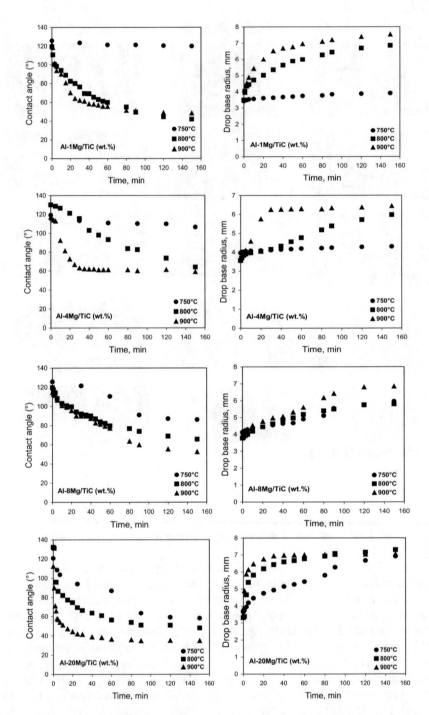

Fig. 2.35 Change of the contact angle and the drop base radius with time for the Al-Mg alloys/TiC systems [112]

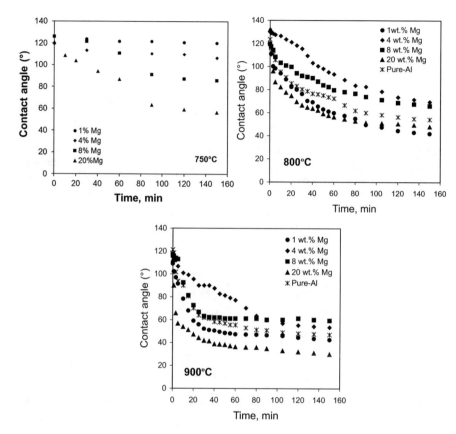

Fig. 2.36 Effect of Mg content in the contact angle for the Al-Mg alloys/TiC systems [112]

Figure 2.37 shows the contact angle curves plotted on a logarithmic time scale. When the contact angle curves were plotted on a logarithmic time scale for 750 and 800 °C, as shown in Fig. 2.37a-b, the contact angles can be seen to progress through the two phases: (I) deoxidation of the Al drop and (II) interfacial reaction wetting phase, where the second phase does not reach the equilibrium; this could be related to the stationary regime that has not yet been reached at this temperature and would need longer period of time to reach the equilibrium. At 900 °C (Fig. 2.37c) the contact angle seems to decrease through three phases: (I) deoxidation of the Al drop, (II) interfacial reaction wetting phase, and (III) equilibrium phase.

According to Fig. 2.37a, the period of the phase I depends of the temperature and the chemical composition of the alloy used. The Al-4 wt.% Mg alloy present the longer phase I. Therefore, the change from the phase I to phase II corresponds to a particular development in the interface reaction.

From Figs. 2.35 and 2.36 it is clearly observed a non-wetting to wetting transition between 750 and 800 °C. However, this transition is time dependent as is shown in

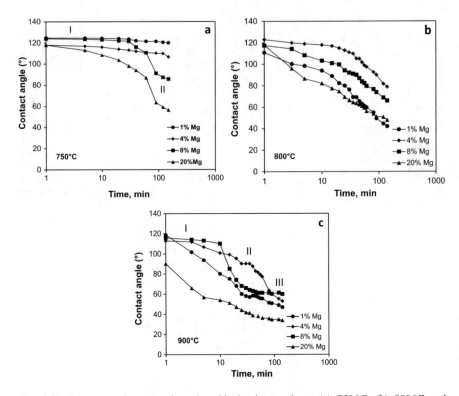

Fig. 2.37 Contact angle curves in a logarithmic time scale at (**a**) 750 °C, (**b**) 800 °C and (**c**) 900 °C [112]

Fig. 2.38. In addition, this transition depends on the chemical composition of the alloy used, showing different behavior depending on the temperature. A non-wetting behavior in all the alloys used at 750 °C after 30 min was observed (Fig. 2.36). However, increasing the temperature at 800 °C with 1 and 20 wt.% of Mg a wetting behavior was observed. Good wetting was obtained at 900 °C after 30 and 60 min.

2.13.3 Thermodynamic Aspects and Magnesium Loss

The wetting of ceramic surfaces by metals and alloys is governed by several chemical and physical processes. This makes theoretical and experimental studies difficult. In particular, the high reactivity of magnesium and magnesium alloys in presence of oxygen is partially described by the reaction:

$$Mg_{(l)} + \tfrac{1}{2}O_{2(g)} = MgO_{(s)}$$
$$\Delta G_{(900°C)} = -473\,kJ$$

(2.19)

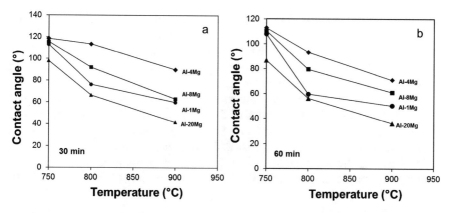

Fig. 2.38 Wetting behavior of Al-Mg/TiC systems in function of the temperature (**a**) after 30 min and (**b**) after 60 min [112]

The reaction between substrate and the melt changes the surface tension during the reaction, because elements with high oxygen affinity as magnesium react with the residual oxygen content in the surrounding atmosphere. Aluminum oxide (Al_2O_3) is stable in pure aluminum but reacts with magnesium in Al-Mg alloys according to the following reactions:

$$3Mg_{(l)} + Al_2O_{3(s)} = 3MgO_{(S)} + 2Al$$
$$\Delta G_{(900°C)} = -123\,kJ \tag{2.20}$$

$$3Mg_{(l)} + 4Al_2O_{3(s)} = 3MgAl_2O_{4(S)} + 2Al$$
$$\Delta G_{(900°C)} = -256\,kJ \tag{2.21}$$

$$MgO_{(s)} + Al_2O_{3(s)} = MgAl_2O_4$$
$$\Delta G_{(900°C)} = -44\,kJ \tag{2.22}$$

If the reaction (2.20) takes place, Mg tends to form a surface oxide in addition to the Al_2O_3 layer, inhibiting a real contact between the TiC and the aluminum. According to thermodynamic prediction reaction (2.22), it will proceed even in a solid state.

While these thermodynamic considerations show the tendency for reaction to occur, it is the kinetics and the extent of reaction which are of practical importance. When the molten aluminum is in real contact with the surface of TiC, spreading is driven solely by chemical reaction at the interface.

Reactions (2.20), (2.21) and (2.22) thermodinamically can be possible to occur, but each one will depend on the temperature and chemical composition of the alloy used. At high magnesium levels, and lower temperatures, MgO may form, while the spinel will form at very low magnesium levels [22, 107, 108].

As observed from Figs. 2.35 and 2.36, the wetting tests of AlMg-alloys on TiC substrates were performed for longer periods of time, and the evaporation of Mg was

observed; even at 750 °C under static argon and atmospheric pressure, part of the Mg
vapor is condensed on the tube wall as Mg or MgO. When the alloy used has high
magnesium content, the evaporation was observed in the top of the drop, leaving a
big quantity of porosity as is shown in Fig. 2.39a. However, this effect was not
observed at low magnesium content (Fig. 2.39b).

The effect of Mg loss on the wetting kinetic was evaluated performing some
interrupted tests at different intervals of time as is shown in Table 2.9. In order to
quantify Mg loss during the wetting process, the magnesium content was analyzed in
drop samples from the interrupted tests at 0, 5, 30, and 120 min.

A more significant magnesium loss for high magnesium content alloys was
observed. A magnesium loss of about 30 wt.% was measured for an Al-20 wt.%

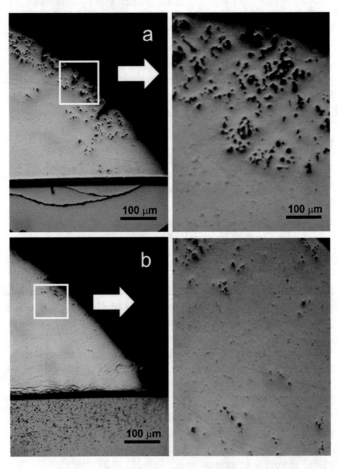

Fig. 2.39 Evaporation effect of Mg in the top of the drop, (**a**) Al-20 wt.% Mg, (**b**) Al-1 wt.% Mg
[112]

Table 2.9 Chemical analysis of the magnesium evaporated at 900 °C

	0 min		5 min		30 min		120 min	
Alloy (wt.%)	Mg	θ	Mg	θ	Mg	θ	Mg	θ
Al-1Mg	1.09	121.3	1.05	93.6	1.03	62	1.02	49
Al-4Mg	4.14	114.8	3.84	106.6	3.76	90	2.09	55.3
Al-8Mg	8.36	117	8.03	113	7.56	63.1	3.73	61.3
Al-20Mg	20.75	112.3	19.97	56.8	13.94	41.8	6.26	35

Table 2.10 Comparison of the magnesium evaporation at 800 and 900 °C

	120 min (800 °C)		120 min (900 °C)	
Alloy (wt.%)	Mg	θ	Mg	θ
Al-1Mg	1.03	49	1.02	44.5
Al-4Mg	3.84	73.6	2.09	55.3
Al-8Mg	5.17	68.8	3.73	61.3
Al-20Mg	13.55	50.8	6.26	35

Mg, after 30 min at 900 °C. However, the contact angle is observed to tend at equilibrium.

The evaporation of magnesium was more significant after long periods of time, which can be minimized using shorter periods of time tests and lower temperatures as is compared in Table 2.10.

Comparing the Mg losses at 800 and 900 °C for Al-20 wt.% Mg after 120 min, Mg losses of about 35 and 70 wt.%, respectively, were observed. The loss of magnesium in function of time for the different Al-Mg alloys employed is shown in the graphic of Fig. 2.40.

It has been suggested through a series of chemical reactions in the system Al-Mg-N_2 that magnesium is retained in a cycle where it is reintroduced to the melt, thus maintaining a low melt surface tension and enhancing wetting [31, 32]. On the other hand, it has been reported that magnesium is consumed due to the formation of the spinel ($MgAl_2O_4$), and a model has been derived to predict the loss of magnesium [108]. It is also been reported that due to its low vapor pressure, magnesium readily volatilizes during holding at the processing temperature. The loss of magnesium from the system has been observed in wetting [31, 87, 88, 106, 112] as well as in infiltration experiments [78, 104, 116, 117].

It is expected that the loss of magnesium from the system does not affect significantly the kinetics of wetting, over all at high magnesium contents, because when magnesium loss was higher, the contact angle tends to be nearly constant. The reactive systems are characterized by pronounced changes of θ with time and strong temperature-dependent variations; these results could make possible to calculate the activation energy for the wetting of Al-Mg alloys/TiC by performing a kinetic study from the spreading of the drop base radius versus time (dr/dt) applying the Arrhenius

Fig. 2.40 (a) Percentage of magnesium loss vs time, (b) remaining magnesium over the time for the different Al-Mg alloys at 900 °C

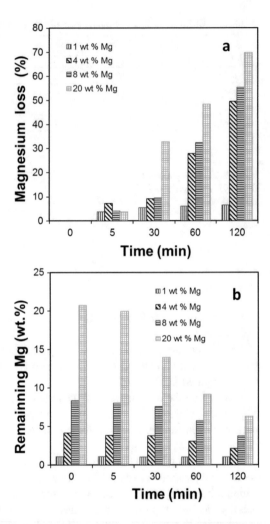

equation as is shown in Fig. 2.41. The high values of activation energy indicates that spreading is not a simple viscosity-controlled phenomenon but is a chemical reaction process [52, 93, 94]. In the present work, the average activation energy obtained for Al-Mg alloys/TiC was 133 kJ/mol, confirming that wetting is driven by reaction at the interface. Table 2.11 shows the activation energy values obtained from the Arrhenius plot of the Fig. 2.41.

2.13.4   Interfacial Characterization

In reactive systems, such as Al-Mg alloys, wetting frequently occurs with extensive chemical reaction and the formation of a new solid compound at the metal/substrate

Fig. 2.41 Arrhenius plot obtained for the different Al-Mg alloys [112]

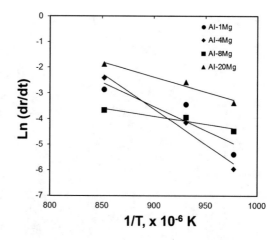

Table 2.11 Activation energy obtained from the Arrhenius plot

Alloy (wt.%)	E_a (kJ/mol)
Al-1 Mg	157
Al-4 Mg	230
Al-8 Mg	52
Al-20Mg	97

interface. Figure 2.42 shows cross sections of the drop-substrate interfaces for Al-Mg alloys/TiC. The thickness of the reaction layer varied within the samples and was discontinuous in nature, particularly for the high Mg contents. The formation of a new interface of about 5–10 μm of thickness is clearly observed for all Al-Mg alloys/TiC systems. Quantitative determinations performed by electron probe microanalysis (EPMA) along different regions of the interface revealed the presence of elementary Al and C with chemical stoichiometries suggesting an Al_4C_3 compound. Other forms of Al_xC_y carbides, with Al-C ratios different to the best-known Al_4C_3 compound, were occasionally detected throughout the interfaces. Traces of Ti were identified in the interface near to the aluminum, which suggested the presence of some titanium aluminide. EDX results showed the interfacial formation of other possible compounds as is shown in Table 2.12. The presence of these elements in the interface suggested that the reaction layer is composed of different products.

Thermodynamic predictions suggest that TiC may be stable in liquid Al above a temperature of 752 °C. Banerji and Reif [84] suggest that TiC is unstable in Al below 1177°C, while Nukami and Flemings [118] pointed out that the reaction between Al, $TiAl_3$, and Al_4C_3 to form TiC starts at 877 °C. Other studies report the formation of the brittle $TiAl_3$ phase can be eliminated entirely using a proper Ti:C molar ratio of 1:1.3 in the Ti-C-Al system, and this phase is always present in the composites when the Ti:C molar ratio is 1:1 [119].

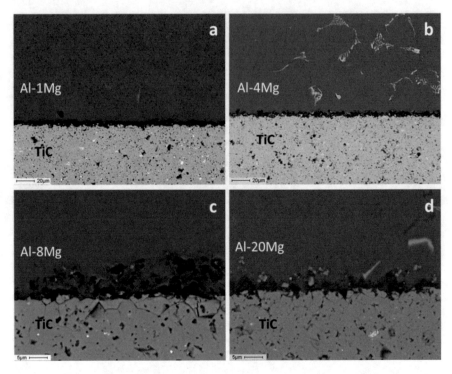

Fig. 2.42 Cross sections of the sessile drop-substrate interface obtained at 900 °C. (**a**) Al-1Mg/TiC, (**b**) Al-4Mg/TiC, (**c**) Al-8Mg/TiC, and (**d**)Al-20Mg/TiC

Table 2.12 Elemental analysis performed at the interface in samples tested at 900 °C

Alloy (wt.%)	Elements detected at the interface	Possible reaction products
Al-1 Mg	Al, C, Mg, O	Al_4C_3, Al_2O_3, $MgAl_2O_4$
Al-4 Mg	Al, C, Mg, O	$MgAl_2O_4$, Al_4C_3, Al_2O_3, MgO
Al-8 Mg	Al, C, Mg, O, Ti	MgO, Al_4C_3, Ti_3AlC, $MgAl_2$
Al-20Mg	Al, C, Mg, O, Ti	MgO, Al_4C_3, Ti_3Al, $TiAl_3$, $MgAl_2$

TiC substrate surface initially flat and smooth after the wetting tests appears rougher because of the dissociation of TiC leading to the formation of Al_4C_3. In a region about 100–1000 μm (depending of the sample and temperature of the test) just below to the drop, the diffusion effect can be observed in the roughness of the substrate leaving a big quantity of microvoids as can be observed in Fig. 2.43. Far of this region, the substrate continues being flat without the presence of microvoids. Thus, it is believed that formation of Al_4C_3 at the interface is greatly affected by the diffusion of carbon when TiC is dissociated, and this is influenced by the temperature and time of the test. The reaction that could rise to the formation of Al_4C_3 and $TiAl_3$ is:

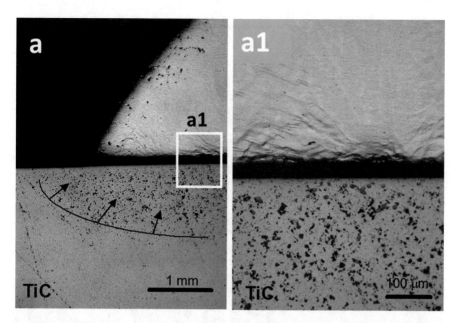

Fig. 2.43 Roughness of the TiC substrate due to dissociation of TiC and diffusion to interface. Al-1Mg/TiC system after 120 min at 900 °C (**a**) at the triple point and (**a1**) magnification at the selected area [112]

$$13Al + 3TiC = 3TiAl_3 + Al_4C_3$$
$$\Delta G_{(750°C)} = -2.14\,kJ$$
$$\Delta G_{(800°C)} = +9.71\,kJ \tag{2.23}$$
$$\Delta G_{(900°C)} = +33.62\,kJ$$

This reaction is thermodynamically possible to occur at temperatures below 752°C, however, it was observed the formation of Al_4C_3 at temperatures of 900 °C, therefore, we suggested that time of the tests and dissociation of TiC and diffusion has effect on the formation of this carbide, also during the cooling the reaction continue in solid state.

If we supposed that substrates have a thin layer of TiO_2 due to the oxidation of titanium, the reaction that could give origin to the formation of $TiAl_3$ and Al_2O_3 is the following:

$$13Al + 3TiO_2 = 3TiAl_3 + 2Al_2O_3$$
$$\Delta G_{(750°C)} = -780\,kJ$$
$$\Delta G_{(800°C)} = -765\,kJ \tag{2.24}$$
$$\Delta G_{(900°C)} = -735\,kJ$$

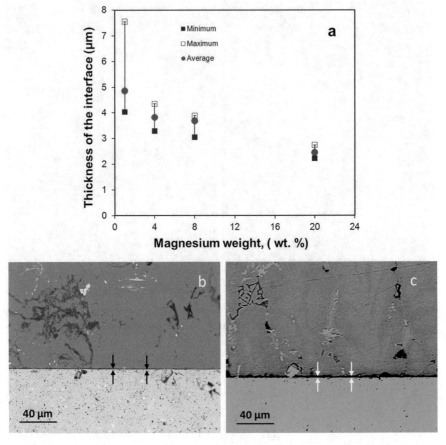

Fig. 2.44 (**a**) Effect of magnesium content in the thickness of the reaction layer at 900 °C, (**b**) Al-20Mg/TiC interface after 5 min, and (**c**) after 30 min [112]

This reaction can be highly possible thermodynamically at all the temperatures studied. Thus, the presence of these interfacial products can be attributed to this reaction.

The precipitation of intermetallic phases delays the formation of Al_4C_3. This was observed at high magnesium contents where the thickness of the reaction layer decreases increasing the magnesium content as is shown in Fig 2.44a, as well as decreasing the temperature and reducing the time of the test as is shown in Fig. 2.44b and c. It is suggested that interfacial reaction thickness increases due to long period of time in the tests influenced by the temperature and the cooling time.

However, the thickness of the reaction products can affect the contact angle if the reaction is intense and leads to a consumption of the reactive species of the liquid alloy and thus to a modification of its composition. In addition, some researchers suggested that $MgAl_2O_4$ spinel improves the wetting [22, 107, 108, 120].

The reaction products will depend on the matrix alloy, being Al_2O_3 for pure aluminum and MgO and $MgAl_2O_4$ for Al-Mg alloys depending of its Mg contain. The spinel reaction can be minimized by using a low Mg-content matrix alloy. Interface reaction can have several undesirable effects. Al_4C_3 dissolves in water, degrading the corrosion behavior. $MgAl_2O_4$ is not expected to affect the corrosion behavior directly, but it will modify the matrix composition. Interfacial reaction products may also modify the mechanical properties of the interface.

The mechanical behavior of the solidified drops under thermal stresses produced during cooling was qualitatively characterized by observing some fractures. In some cases (especially at high Mg contents), a cohesive fracture (bulk rupture path in TiC near the interface) was observed corresponding to a strong interface as is shown in Fig. 2.45.

The work of adhesion of the metal-ceramic couples was calculated in function of the contact angle (θ) and surface tension (γ_{LV}), using the Eq. (2.4), where the values of γ_{LV} were calculated using data in reference [91]. Table 2.13 shows the results obtained. The work of adhesion is a measure of the strength of the joining between the ceramic (TiC) and metal.

Fig. 2.45 Typical cohesive fractures observed, Al-8Mg/TiC at 900 °C

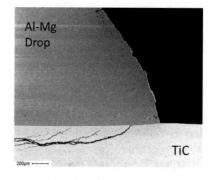

Table 2.13 Adhesion work evaluated in function of the surface tension and contact angle

System	Adhesion work (mJ/m^2)[a]		
	750 °C	800 °C	900 °C
Al-1Mg/TiC	455	1359	1457
Al-4Mg/TiC	604	1151	1250
Al-8Mg/TiC	939	1201	1356
Al-20Mg/TiC	1289	1367	1465
Pure-Mg/TiC	416	435	904
Pure-Al/TiC	–	1317	1405

[a]The contact angle used was the value obtained after 120, 60, and 30 min for the temperature of 750, 800, and 900 °C, respectively. Due to after this time, the evaporation of Mg is too much. In addition, at these times the contact angle is nearly constant.

When there is a good wetting, generally $W_a \geq 2\gamma_{LV}$. This means that the adhesion energy between the ceramic and the melt should be more than twice the surface tension of the liquid. The W_a for Mg/TiC system calculated at 800 °C was 435 mJ/m^2 . Considering that $\gamma_{LV(Mg)} = 538$ mJ/m^2, this means about 40% of the cohesion work of liquid Mg, which suggests that Mg/TiC interface will be weak energetically. In this nonreactive system, the adhesion could be attributed to van der Waals forces mainly. In the case of Al-1Mg/TiC at the same temperature, the work of adhesion represents 90% of the cohesion work of liquid Al indicating a strong interface. Thus, adhesion in Al-1Mg/TiC system cannot be interpreted in terms of weak interactions of van der Waals but can be interpreted in terms of strong chemical interactions, such as covalent bond.

Wetting can be obtained not only in systems featuring strong liquid/solid interactions (ionic, covalent, metallic, or some mixture of them) but also in systems featuring weak liquid/solid interactions (for instance, van der Waals forces) provided that liquid/liquid interactions are weak too.

From the above, it is obvious that good wetting does not necessarily mean a thermodynamically strong interface [110]. Good wetting means that the interfacial bond is energetically nearly as strong as the cohesion bond of the liquid [111].

2.14 Wettability of TiC by Commercial Aluminum Alloys

The wetting of molten metals on ceramic surfaces is of crucial importance in liquid-state processing of metal-ceramic composites. In general, ceramics are not readily wetted by liquid metals at the usual melt temperatures. Consequently, wettability enhancing procedures are applied to assist wettability, such as the addition of suitable alloying elements in the melt [9, 18, 20, 76, 106, 112, 121]. The effect of alloying elements on the wettability of TiC by commercial aluminum alloys (1010, 2024, 6061, and 7075) was investigated at 900 °C using a sessile drop technique [122]. Changes in contact angles and spreading of the drops were determined as a function of temperature and time in order to establish the effect of the alloying elements on the wetting behavior.

2.14.1 Substrate Preparation

TiC substrates were prepared from c.a.s. grade TiC powder (H. C. Starck, Germany), having a particle size distribution from 0.3 to 3 μm. The chemical composition of the powders is shown in Table 2.14.

Table 2.14 Chemical composition of TiC powder

Ti	C(total)	C(free)	O	N
79.7%	19.26%	0.1%	0.49%	0.04%

   Dense TiC substrates were obtained by hot-pressing 5 g of the as-received material in a 2.5-cm-diameter graphite die as is shown in Fig. 2.12. The used conditions were 30 min at 1800 °C using a 30 MPa load under ~20 Pa vacuum. Density of the sintered substrates was determined by the Archimedes method (ASTM C 373-88), reaching 96.8%. Scanning electron microscopy (SEM) observation showed residual porosity (closed pores) with less than 1 μm. The contact angle needs to be measured on smooth surfaces; therefore, the discs were polished to a mirrorlike finish with 1 μm diamond paste. Roughness measurements performed by atomic force microscopy (AFM) revealed mean roughness values, R_a, from 2.12 to 2.76 nm. The surface roughness factor, R_w, defined as the ratio of the true to apparent surfaces areas, was 0.032%. Therefore, hysteresis of the contact angle caused by roughness was minimized.

2.14.2 Wetting Experimental Conditions

Wetting experiments were carried out using a sessile drop technique. The used experimental equipment was shown in Fig. 2.13, which consists of a quartz tube furnace, designed specifically for sessile drop tests and connected to a high vacuum system operating at an absolute pressure of 10^{-4}–10^{-5} Torr. The tests were performed at 900 °C for all the alloys, under vacuum and argon atmospheres. For vacuum operation, the local O_2 partial pressure was reduced by flushing argon and placing a Ti sponge getter in the vicinity of the hot zone. For wetting experiments in argon, the system was first evacuated and then backfilled with ultrahigh purity argon (99.99%) at 0.35 kPa.

   Table 2.15 shows the chemical composition for the aluminum alloys (1010, 2024, 6061, and 7075) used. Metal samples were cleaned in hydrofluoric acid and then in acetone prior to testing. Approximately 0.8 g of metal was placed on the TiC substrates in the furnace before heating. Once the desired temperature was stabilized, the Al alloy/TiC assembly was pushed into the tube hot zone. At the time the metal drop was well shaped, contact angle changes with time were followed until a steady state was reached and changes in θ were not appreciable. The contact angle and spreading radius were recorded photographically at various time intervals, and accuracy is measured directly from the image of the drop section using computational tools.

Table 2.15 Chemical composition of commercial alloys used

Alloy	Si	Fe	Cu	Mn	Mg	Cr	Ni	Zn	% alloying elements
Al-1010	0.20	0.65	0.005	0.002	0.014	0.012	0.009	0.002	0.89
Al-2024	0.35	0.36	4.46	1.08	1.86	0.017	–	0.047	8.17
Al-6061	0.61	0.40	0.25	0.06	0.82	0.18	<0.009	0.07	2.39
Al-7075	0.25	0.60	2.21	0.26	2.32	0.22	–	5.52	11.38

After completion of the experiment, the samples were immediately removed from the hot zone inducing fast cooling. The metal-ceramic interfaces were analyzed by electron probe microanalysis (EPMA) equipped with wavelength dispersive spectroscopy facilities (WDS) for quantitative compositional analysis. Conducting X-ray mapping and line scan distribution of the elements present, as well as quantitative microanalysis for phase identification, made possible direct correlations between structure and composition.

2.14.3   Wetting and Spreading Kinetic

The variation of the contact angle of the Al alloys used and TiC are shown in Fig. 2.46. Figure 2.46a shows the wettability of TiC by Al alloys under vacuum and Fig. 2.46b shows the wettability of TiC by Al alloys under argon at 900 °C. Generally speaking, the wetting increased in the order 6061 < 7075 < 2024 < 1010 for both vacuum and argon atmospheres.

Wetting of pure aluminum on TiC occurred at 900 °C with a contact angle lower than 60°. After equilibrium was reached, the contact angles do not differ significantly from the values reported by Muscat and Drew [16]. However, Kononenko [74] and Frumin [75] reported a non-wetting to wetting transition at ~1050°C. Because wetting conditions were attained at 900 °C, it is suggested that the temperature for wetting transition in the Al/TiC system must be lower than that proposed by Kononenko [74] and Frumin [75], being lower than 900 °C as was reported elsewhere [78, 87].

The effect of atmosphere on the wetting behavior was most evident for the 6061 and 7075 alloys, with 6061 exhibiting the slowest spreading kinetics. When testing these alloys under vacuum conditions (Fig. 2.46a), evaporation of alloying elements was observed; a metallic film was deposited in the wall of the quartz tube, mostly near to the cooler parts of the tube. While the 6061 alloys showed a very small

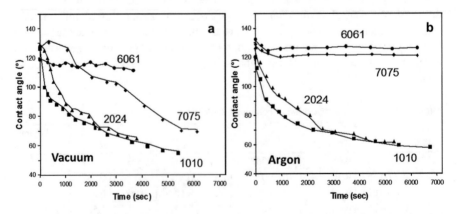

Fig. 2.46 Wettability of TiC by Al alloys, (**a**) under vacuum and (**b**) argon at 900 °C [122]

variation of θ with time, the contact angle of the 7075 decreased slowly until reaching a wetting angle close to the equilibrium.

The effect of the vapor pressure and the continuous evaporation of the low boiling point elements such as Zn (907 °C) and Mg (1120 °C) decreased to some extent the drop volume. A fluctuating behavior of the wetting curves for the 6061 and 7075 alloys in vacuum was observed (Fig. 2.46a). Meanwhile under argon (Fig. 2.46b) a non-wetting behavior for Al-6061 and Al-7075 was observed.

A different behavior was observed when the 6061 and 7075 alloys were tested under an argon atmosphere. After an initial slight decrease of θ, the contact angles remained nearly constant for 2 h and steady-state values greater than 118° were obtained for both alloys. The use of argon certainly minimized evaporation of the alloying elements from the drop.

The wetting curves of Al-1010 and Al-2024 exhibited a characteristic behavior of non-equilibrium systems, showing a sharp decrease of θ in the first seconds followed by a transition region before a steady state was achieved. The main changes in θ took place within the first 10 min. The wetting curves of 1010 aluminum were similar irrespective of the atmosphere, and a slightly lower constant θ was observed under vacuum conditions. This is attributed to the more stable oxide film retained in argon with the consequent delay in spreading of the drop. When the liquid alloy is plainly in contact with the surface of TiC, spreading is driven solely by chemical reaction at the interface [16], which reduces the interfacial tension. Thus, the decrease of the contact angle is proportional to the rate of chemical reaction, which is slightly influenced by the presence of the alloying elements.

Because the spreading kinetics of Al-1010 was faster than the alloys, the contact angle rapidly approached the steady-state value. Spreading of Al alloys on TiC in vacuum and argon is shown in Fig. 2.47. The behavior of the drop radius curves shown in Fig. 2.47 suggests that the alloying elements modify the nature of the oxide layer on the drop; thus evaporation becomes a more complex process which results in delayed spreading. Eustathopoulos [44] suggested that only those alloying elements affecting the compactness of the oxide layer have a significant effect on the contact angle. It is known that Mg improves wettability because its rapid evaporation

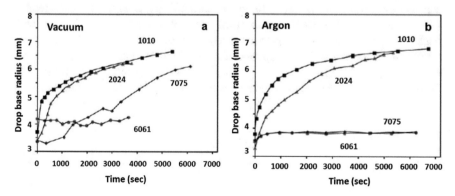

Fig. 2.47 Spreading of Al alloys on TiC, (**a**) under vacuum and (**b**) argon at 900 °C [122]

under vacuum damages the alumina film protecting the metal drop. Thus, in the present experiments, the high Mg content in the 2024 and 7075 alloys could lead to the improved wetting behavior under vacuum. It is reported that during the wetting of alumina by a 2.5% Mg containing Al alloy, magnesium volatilized completely at 700 °C after 3 min following melting, thus improving wettability [30]. On the other hand, Mg is also highly reactive, and the free energy of formation of its oxide is more negative than that of the oxides of aluminum. This suggests that Mg reduces the aluminum oxide rupturing the oxide film sufficiently to facilitate diffusion and wetting. It is reported that spinel ($MgAl_2O_4$) occurs at low magnesium contents while MgO forms by direct contact with Al-Mg eutectic [123].

As was expected, evaporation of the alloying elements contained in 6061 and 7075 alloys decreased in argon, increasing the tenacity of the oxide layer due to their incorporation into the film [44]. This led to a stable and compacted layer containing complex oxides which are more stable under the processing conditions. This film is difficult to break up even after extended periods of time, hence delaying wetting. Results on the evaporation of commercial aluminum alloys have been reported using thermogravimetric analyses in the processing of Al/TiC composites by the current authors [116, 117]. Evaporation of Zn and Mg under vacuum conditions was noticeable by the low boiling point elements Zn (907 °C) and Mg (1120 °C).

In the case of vacuum conditions, the evaporation of Zn (7075) and Mg (2024) contributed to the rupture of the oxide film covering the aluminum drop and thereby improving wetting. Evaporation from the 6061 was more gradual and less excessive, probably because of the lower Mg content (0.82%) in this alloy.

2.14.4 Interfacial Characterization

In reactive systems, the change in contact angle generally implies that chemical reactions occurred at the Al alloy/TiC interface. By EDX microanalyses it was evaluated the elements present at the interface and by means of stoichiometry analysis was estimated the Al_4C_3 compound at the interface in all the specimens in different amounts. This carbide was mainly present in Al-1010/TiC system, with thickness up to 6–7 µm. Similarly than for Al_4C_3, EDX results showed the interfacial formation of other Ti-Al-C compounds in the 1010/TiC and 7075/TiC couples, which best fits the formation of $AlTiC_2$ and Al_5TiC_4, respectively. Table 2.16 shows the chemical elemental composition determined at interface. Other forms of

Table 2.16 Possible formation phases identified by stoichiometry analysis of EDX (at.%)	Composition (at.%)			
	Al	C	Ti	Estimated phase
	56.5	42.2	1.0	Al_4C_3
	25.7	47.9	26.4	$AlTiC_2$
	50.8	39.4	9.6	Al_5TiC_4

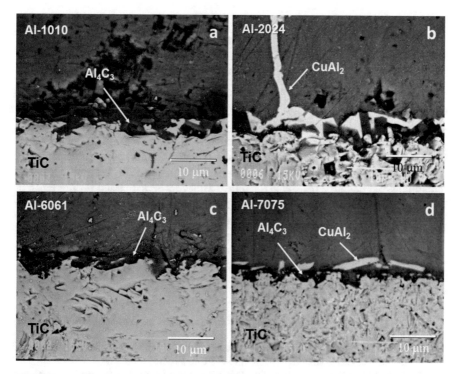

Fig. 2.48 Al/TiC interfaces (**a**) Al-1010, (**b**) Al-2024, (**c**) Al-6061, and (**d**) Al-7075 [122]

Al_xC_y carbides, with Al-C ratios different to the best-known Al_4C_3 compound, were detected throughout the interfaces.

Figure 2.48 shows typical interfaces found in samples tested under vacuum. All the samples present interfacial reaction products. The ceramic surface initially flat and smooth appears rougher because of the dissociation of TiC leading to the formation of Al_4C_3. The thickness of the reaction layer varied within the samples and was discontinuous, particularly for the 7075/TiC and 6061/TiC systems, which exhibited poor wetting in comparison with 1010 and 2024. Intermetallic phases like $CuAl_2$ were observed close to the interface in the 2024/TiC and 7075/TiC.

Examination of drop sections showed that the precipitation of alloyed phases in the ceramic surface decreases the amount of the undesirable Al_4C_3 at the metal-ceramic interface. When the chemical reaction takes place at the solid-liquid interface, TiC dissociates and Al_4C_3 forms. However, no traces of Ti were identified in the aluminum nor was the presence of $TiAl_3$ or any other titanium aluminide observed. As magnesium does not form magnesium carbide, no reaction-aided wetting at the interface involving magnesium occurred. However, the precipitation of $CuAl_2$ and Mg_2Si was usually observed at the interface.

Froumin et al. [100] reported that in the presence of TiC, alloying with Cu decreases the solubility of Ti in molten Al. Hence this reduces the amount of titanium being transferred into the melt and thereby lowering wetting of the drop.

Table 2.17 Work of adhesion calculated in function of the surface tension (γ_{LV}) and the contact angle (θ)

Al alloy	Atmosphere	Adhesion work (mJ/m^2)
Al-1010	Argon	1190
Al-1010	Vacuum	1224
Al-2024	Argon	1207
Al-2024	Vacuum	1254

The work of adhesion of the 1010/TiC and 2024/TiC systems is calculated from the surface tension data given by Orkasov [124] and contact angle values corresponding to 1 h holding time. Table 2.17 show the adhesion work results for Al-1010 and Al-2024 indicate that adhesion was slightly higher under vacuum conditions, particularly for the 2024 alloy. This can be explained on the basis of the drop being less oxidized, but, above all, it indicates the contribution of the copper precipitates near to the interface on the energy of adhesion.

References

1. Gibbs JW (1878) On the equilibrium of heterogeneous substances. Trans Conn Acad 3:343–524
2. Jhonson RE (1959) Conflicts between Gibbsian thermodynamics and recent treatments of interfacial energies in solid-liquid-vapor. J Phys Chem 63:1655–1658
3. Naidich JV (1981) In: Cadenhead DA, Danielli JF (eds) Progress in surface and membrane science, vol 14. Academic Press, Cambridge, pp 353–484
4. Gallois BM (1997) Overview: wetting in nonreactive liquid metal-oxide systems. JOM 49 (6):48–51
5. Kaptay G (1996) Interfacial phenomena during melt processing of ceramic particle-reinforced metal matrix composites. Mater Sci Forum 215–216:459–466
6. Ruhle M (1996) Structure and composition of metal/ceramic interfaces. J Eur Ceram Soc 16 (3):353–365
7. Savov L, Heller HP, Janke D (1997) Wettability of solids by molten metals and alloys. Metall 51(9):475–486
8. Dalgleish BJ, Saiz E, Tomsia AP, Cannon RM, Ritchie RO (1994) Interface formation and strength in ceramic/metal systems. Scr Metall Mater 31(8):1109–1114
9. Delannay F, Froyen L, Deruyttere A (1987) The wetting of solids by molten metals and its relation to the preparation of metal-matrix composites. J Mater Sci 22(1):1–16
10. Nowok JW (1994) Analysis of atomic diffusion in liquid metals at melting temperatures in capillary like Media-2. Acta Metall Mater 42(12):4025–4028
11. Eustathopoulus N (1998) Dynamics of wetting in reactive metal/ceramic systems. Mater Sci Eng 249A(1):176–183
12. Li JG (1994) Wetting of ceramic materials by liquid Si, Al and other metallic melts containing Ti and other reactive elements. Rev Ceram Int 20(6):391–412
13. Aksay IA, Hoge CE, Pask JA (1974) Wetting under chemical equilibrium and non-equilibrium conditions. J Phys Chem 78(12):1178–1183
14. Naidich YV, Taranets NY (1988) Wettability of aluminum nitride by tin-aluminum melts. J Mater Sci 33:393–397
15. Samsonov GV, Panasyuk AD, Kozina GK (1968) Wetting of refractory carbides with liquid metals. Porosk Metall 71(11):42–48

16. Muscat D, Drew RAL (1994) Modeling the infiltration kinetics of molten aluminum into porous titanium carbide. Metall Mater Trans 25A(11):2357–2370

17. Muscat D, Harris RL, Drew RAL (1994) The effect of pore size on the infiltration kinetics of aluminum in TiC preforms. Acta Metall Mater 42(12):4155–4163

18. Banerji A, Rohatgi PK, Reif W (1984) Role of the wettability in the preparation of metal-matrix composites (a review). Metall 38:656–661

19. León CA, Drew RAL (2000) Preparation of nickel-coated powders as precursors to reinforce MMC's. J Mater Sci 35(19):4763–4768

20. León CA, Bourassa AM, Drew RAL (2000) Processing of aluminum matrix composites by electroless plating and melt infiltration. Adv Technol Mater Mater Process J 2(2):96–106

21. Hatch JE (1984) ALUMINUM properties and physical metallurgy. ASM International, Geauga

22. Pai BC, Ramani G, Pillai RM, Satyanarayana KG (1995) Review: Role of magnesium in cast aluminum alloy matrix composites. J Mater Sci 30:1903–1911

23. Mcevoy AJ, Williams RH, Higginbotham IG (1976) Metal/non-metal interfaces. The wetting of magnesium oxide by aluminum and other metals. J Mater Sci 11:297–302

24. Brewer L, Searcy AW (1951) The gaseous species of the Al-Al_2O_3 system. J Am Chem Soc 73:5308–5314

25. Brennan JJ, Pask JA (1968) Effect of nature of surfaces on wetting of saphire by liquid aluminum. J Am Ceram Soc 51(10):569–573

26. Porter RF, Schissel P, Inghram MG (1955) A mass spectrometric study of gaseous species in the Al-Al_2O_3 system. J Chem Phys 23(2):339–342

27. Rao YK (1985) Stoichiometry and thermodynamics of metallurgical processes. CBLS Publishers, Marietta

28. López Morelos VH (2000) Mojabilidad del TiC por el Aluminio y sus Aleaciones. Thesis of Master degree, IIM-UMSNH, Morelia Mich., México

29. Madeleno U, Liu H, Shinoda T, Mishima Y, Suzuki T (1990) Compatibility between alumina fibres and aluminum. J Mater Sci 25:3273–3280

30. Lijun Z, Jimbo W, Jiting Q, Qiu N (1989) An investigation on wetting behavior and interfacial reactions of aluminum α-Alumina system. In: Lin RY et al (eds) Proceeding of interfaces in metal-ceramics composites. TMS, Warrendale, pp 213–226

31. Pech-Canul MI, Katz RN, Makhlouf MM (2000) Optimum parameters for wetting silicon carbide by aluminum alloys. Metall Mater Trans 31A:565–573

32. Pech-Canul MI, Katz RN, Makhlouf MM (2000) The combined role of nitrogen and magnesium in wetting SiC by aluminum alloys. In: Memoria XXII Congreso Internacional de Metalurgia y Materiales, Saltillo Coah., México, pp 232–241

33. García-Cordovilla C, Louis E, Pamies A (1986) The surface tension of liquid pure aluminium and aluminium-magnesium alloy. J Mater Sci 31(21):2787–2792

34. Goicoechea J, García-Cordovilla C, Louis E, Pamies A (1992) Surface tension of binary and ternary aluminum alloys of the systems Al-Si-Mg and Al-Zn-Mg. J Mater Sci 27:5247–5252

35. Narciso J, Alonso A, Pamies A, García-Cordovilla C, Louis E (1994) Wettability of binary and ternary alloys of the system Al-Si-Mg with SiC particulates. Scr Metall Mater 31 (11):1495–1500

36. Manning CR, Gurganus TB (1969) Wetting of binary aluminum alloys in contact with Be, B_4C, and graphite. J Am Ceram Soc 52(3):115–118

37. Pai BC, Ray S, Prabhakar KV, Rohatgi PK (1976) Fabrication of aluminum-alumina/magnesia/particulate composites in foundries using magnesium additions to the melts. Mater Sci Eng 24:31–44

38. Dean WA (1967) In: Horn V (ed) Aluminum, vol 1. ASM Pub, Metals Park, Ohio, p 163

39. Suresh S, Mortensen A, Needleman A (1993) Fundamentals of metal matrix composites. Butterworth-Heinemann, Boston

40. Banerji A, Rohatgi K (1982) Cast aluminum alloy containing dispersions of TiO_2 and ZrO_2 particles. J Mater Sci 17(2):335–342

41. Laurent V, Chatain D, Eustathopoulos N (1991) Wettability of SiO$_2$ and oxidized SiC by aluminum. Mater Sci Eng 135:89–94
42. Bardal A (1992) Wettability and interfacial reaction products in the AlSiMg surface-oxidized SiC system. Mater Sci Eng 159A:119–125
43. Eustathopoulos N, Drevet B (1998) Determination of the nature of metal-oxide interfacial interactions from Sessile drop data. Mater Sci Eng A 249(1):176–183
44. Eustathopoulos N, Joud JC, Desre P, Hicter JM (1974) The wetting of carbon by aluminum and aluminum alloys. J Mater Sci 9(8):1233–1242
45. Pique D, Coudurier L, Eustathopoulos N (1981) Adsorption du cuivre a l'interface entre Fe solide et Ag liquide a 1100 °C. Scr Metall 15(2):165–170
46. Humenik M, Kingery WD (1954) Metal-ceramic interactions III: surface tension and wettability of metal-ceramic systems. J Am Ceram Soc 37(1):18–23
47. Wenzel RN (1936) Resistance of solid surfaces to wetting by water. Ind Eng Chem 28(8):988–994
48. Nakae H, Inui R, Hirata Y, Saito H (1998) Effects of surface roughness on wettability. Acta Metall Mater 46(7):2313–2318
49. Eustathopoulos N (1998) Dynamics of wetting in reactive metal/ceramics systems. Acta Mater 46(7):2319–2327
50. Dezellus O, Eustathopoulos N (2010) Fundamental issues of reactive wetting by liquid metals. J Mater Sci 45:4256–4264
51. Dezellus O, Eustathopoulos N (1999) The role of Van der Waals interactions on wetting and adhesion in metal/carbon systems. Scr Mater 40(11):1283–1288
52. Dezellus O, Hodaj F, Eustathopoulos N (2002) Chemical reaction-limited spreading: the triple line velocity versus contact angle relation. Acta Mater 50:4741–4753
53. Dezellus O, Hodaj F, Eustathopoulos N (2003) Progress in modelling of chemical-reaction limited wetting. J Eur Ceram Soc 23(15):2797–2803
54. Dezellus O, Hodaj F, Mortensen A, Eustathopoulos N (2001) Diffusion-limited reactive wetting. Spreading of Cu-Sn-Ti alloys on vitreous carbon. Scr Mater 44:2543–2549
55. Mortensen A, Drevet B, Eustathopoulos N (1997) Kinetic of diffusion-limited spreading of sessile drops in reactive wetting. Scr Mater 36(6):645–651
56. Frage N, Froumin N, Dariel MP (2002) Wetting of TiC by non-reactive liquid metals. Acta Mater 50(2):237–245
57. Asthana R, Sobezak N (2000) Wettability, spreading and interfacial phenomena in high temperature coatings. JOM 52(1):1–19
58. Starov VM, Velarde MG, Radke CJ (2007) Wetting and spreading dynamics, vol 138. CRC Press, Boca Raton
59. Saiz E, Tomsia AP, Cannon RM (1998) Ridging effects on wetting and spreading of liquids on solids. Acta Mater 46(7):2349–2361
60. Lam CNC, Wu R, Lia D, Hair ML, Neumann AW (2002) Study of the advancing and receding contact angles: liquid sorption as a cause of contact angle hysteresis. Adv Colloid Interface Sci 96:169–191
61. Eick JD, Good RJ, Neumann AW (1975) Thermodynamics of contact angles. II. Rough solid surfaces. J Colloid Interface Sci 53(2):235–248
62. Oliver JF, Huh C, Mason SG (1980) An experimental study of some effects of solid surface roughness on wetting. Colloids Surf 1:79
63. Oliver JF, Mason SG (1980) Liquid spreading on rough metal surfaces. J Mater Sci 15(2):431–437
64. Neumann AW, Good RJ (1972) Thermodynamics of contact angles. I. Heterogeneous solid surfaces. J Colloid Interface Sci 38:341–358
65. Marmur A (1997) Line tension and the intrinsic contact angle in solid–liquid–fluid systems. J Colloid Interface Sci 186(2):462–466
66. Decker EL, Garoff S (1997) Contact line structure and dynamics on surfaces with contact angle hysteresis. Langmuir 13(23):6321–6332

67. Fadeev AY, McCarthy TJ (1999) Binary monolayer mixtures: modification of nanopores in silicon supported tris (trimethylsiloxy) silyl monolayers. Langmuir 15:7238–7243
68. Fadeev AY, McCarthy TJ (1999) Trialkylsilane monolayers covalently attached to silicon surfaces: wettability studies indicating that molecular topography contributes to contact angle hysteresis. Langmuir 15:3759–3766
69. Youngblood JP, McCarthy TJ (1999) Ultrahydrophobic polymer surfaces prepared by simultaneous ablation of polypropylene and sputtering of poly (tetrafluoroethylene) using radio frequency plasma. Macromolecules 32:6800–6806
70. Sedev RV, Petrov JG, Neumann AW (1996) Effect of swelling of a polymer surface on advancing and receding contact angles. J Colloid Interface Sci 180:36–42
71. Lam CNC, Wu R, Li D, Hair ML, Neumann AW (2002) Study of the advancing and receding contact angles: liquid sorption as a cause of contact angle hysteresis. J Colloid Interface Sci 96:169–191
72. Jha AK, Prasad SV, Upadhyaya GS (1990) In: Bhagat RB (ed) Metal & ceramic matrix composites. CRC Press, Boca Raton, pp 127–135
73. Rhee SK (1970) Wetting of ceramics by liquid aluminum. J Am Ceram Soc 53(7):386–389
74. Kononenko VY, Shvejkin GP, Sukhman AL, Lomovtsev VI, Mitrofanov BV (1976) Chemical compatibility of titanium carbide with aluminum, gallium, and indium melts. Poroshk Metall 9:48–52
75. Frumin N, Frage N, Polak M, Dariel MP (1997) Wettability and phase formation in the TiC$_x$/Al system. Scr Mater 37(8):1263–1267
76. Asthana R, Tewari SN (1993) Interfacial and capillary phenomena in solidification processing of metal-matrix composites. Compos Manuf 4(1):3–25
77. Kaptay G, Bader E, Bolyan L (2000) Interfacial forces and energies relevant to production of metal matrix composites. Mater Sci Forum 329–330:151–156
78. Contreras A, López VH, León CA, Drew RAL, Bedolla E (2001) The relation between wetting and infiltration behavior in the Al-1010/TiC and Al-2024/TiC systems. Adv Technol Mater Mater Process 3(1):33–40
79. Ferro AC, Derby B (1995) Wetting behavior in the Al-Si/SiC system: interface reactions and solubility effects. Acta Metall Mater 43(8):3061–3073
80. Lin Q, Shen P, Yang L, Jin S, Jiang Q (2011) Wetting of TiC by molten Al at 1123–1323 K. Acta Mater 59:1898–1911
81. Xiao P, Derby B (1996) Wetting of titanium nitride and titanium carbide by liquid metals. Acta Mater 44(1):307–314
82. Schuster CJ, Nowotny H, Vaccaro C (1980) The ternary systems: Cr-Al-C, V-Al-C, and Ti-C-Al and the behavior of H-phases (M$_2$AlC). J Solid State Chem 32:213–219
83. Iseki T, Kameda T, Maruyama T (1983) Some properties of sintered Al$_4$C$_3$. J Mater Sci Lett 2:675–676
84. Banerji A, Reif W (1986) Development of Al-Ti-C grain refiners containing TiC. Metall Trans 17A:2127–2137
85. Fine ME, Conley JG (1990) On the free energy of formation of TiC and Al$_4$C$_3$. Metall Trans 21A:2609–2610
86. Yokokawa H, Sakai N, Kawada T, Dakiya M (1991) Chemical potential diagram of Al-Ti-C System: Al$_4$C$_3$ formation on TiC formed in Al-Ti liquids containing carbon. Metall Trans 22A:3075–3076
87. Contreras A, Leon CA, Drew RAL, Bedolla E (2003) Wettability and spreading kinetics of Al and Mg on TiC. Scr Mater 48:1625–1630
88. Contreras A (2002) Fabricación y estudio cinético de materiales compuestos de matriz metálica Al-Cu$_x$ y Al-Mg$_x$ reforzados con TiC: Mojabilidad e infiltración. Thesis, Universidad Nacional Autónoma de México
89. Laurent V, Chatain D, Chatillon C, Eustathopoulos N (1998) Wettability of monocrystalline alumina by aluminum between its melting point and 1273K. Acta Metall 36(7):1797–1803

90. Brennan JJ, Pask JA (1968) Effect of composition on glass-metal interface reactions and adherence. J Am Ceram Soc 56(2):58–62
91. Keene BJ (1993) Review of data for the surface tension of pure metals. Int Mater Rev 38 (4):157–192
92. Muscat D (1993) Titanium carbide/Aluminum composites by melt infiltration. Thesis, Department of Mining and Metallurgical Engineering, McGill University, pp 48–51
93. Kumar G, Narayan K (2007) Review of non-reactive and reactive wetting of liquids on surfaces. Adv Colloid Interface Sci 133:61–89
94. Toy C, Scott WD (1997) Wetting and spreading of molten aluminium against AlN surfaces. J Mater Sci 32:3243–3248
95. Narayan K, Fernandes P (2007) Determination of wetting behavior, spread activation energy, and quench severity of bioquenchants. Metall Mater Trans B 38:631–640
96. Contreras A (2007) Wetting of TiC by Al–Cu alloys and interfacial characterization. J Colloid Interface Sci 311:159–170
97. Li L, Wong YS, Fuh JYH, Lu L (2001) Effect of TiC in copper–tungsten electrodes on EDM performance. J Mater Process Technol 113:563–567
98. Leong CC, Lu L, Fuh JYH, Wong YS (2002) In-situ formation of copper matrix composites by laser sintering. Mater Sci Eng A 338:81–88
99. Akhtar F, Javid-Askari S, Ali-Shah K, Du X, Guo S (2009) Microstructure, mechanical properties, electrical conductivity and wear behavior of high volume TiC reinforced Cu-matrix composites. Mater Charact 60:327–336
100. Froumin N, Frage N, Polak M, Dariel MP (2000) Wetting phenomena in the TiC/(Cu-Al) system. Acta Mater 48:1435–1441
101. Mortimer DA, Nicholas M (1973) The wetting of carbon and carbides by copper alloys. J Mater Sci 8:640–648
102. Zarrinfar N, Kennedy AR, Shipway PH (2004) Reaction synthesis of Cu-TiC$_x$ master-alloys for the production of copper-based composites. Scr Mater 50:949–952
103. Zarrinfar N, Shipway PH, Kennedy AR, Saidi A (2002) Carbide stoichiometry in TiC$_x$ and Cu-TiC$_x$ produced by self-propagating high-temperature synthesis. Scr Mater 46:121–126
104. Contreras A, Albiter A, Bedolla E, Perez R (2004) Processing and characterization of Al-Cu and Al-Mg base composites reinforced with TiC. Adv Eng Mater 6:767–775
105. Shoutens JE (1992) Some theoretical considerations of the surface tension of liquid metals for metal matrix composites. J Mater Sci 24:2681–2686
106. Aguilar EA, Leon CA, Contreras A, Lopez VH, Drew RAL, Bedolla E (2002) Wettability and phase formation in TiC/Al-alloys assemblies. Compos Part A 33:1425–1428
107. Lloyd DJ (1994) Particle reinforced aluminium and magnesium matrix composites. Int Mater Rev 39:1–24
108. McLeod AD, Gabryel CM (1992) Kinetic of the grow spinel MgAl$_2$O$_4$ on alumina particulate in aluminum alloys containing magnesium. Metall Trans A 23:1279–1283
109. Saiz E, Tomsia AP (1998) Kinetics of metal-ceramic composite formation by reactive penetration of silicates with molten aluminum. J Am Ceram Soc 81(9):2381–2393
110. Yosomiya R, Morimoto K, Nakajima A, Ikada Y, Suzuki T (eds) (1990) Adhesion and bonding in composites. Marcel Dekker, New York, p 23
111. Eustathopoulos N, Nicholas MG, Drevet B (1999) In: Cahn RW (ed) Wettability at high temperatures, Pergamon materials series, vol 3. Elsevier Science & Technology, Oxford, p 45
112. Contreras A, Bedolla E, Perez R (2004) Interfacial phenomena in wettability of TiC by Al–Mg alloys. Acta Mater 52:985–994
113. Yoshimi N, Nakae H, Fujii H (1990) A new approach to estimating wetting in reaction system. Mater Trans JIM 31(2):141–147
114. Nakae H, Fujii H, Sato K (1992) Reactive wetting of ceramics by liquid metals. Mater Trans JIM 33:400–406
115. Fujii H, Nakae H (1990) Three wetting phases in the chemically reactive MgO/Al system. ISIJ Int 30(12):1114–1118

116. Contreras A, Salazar M, León CA, Drew RAL, Bedolla E (2000) Kinetic study of the infiltration of aluminum alloys into TiC. Mater Manuf Process 15(2):163–182
117. Contreras A, Albiter A, Perez R (2004) Microstructural properties of the Al–Mg$_x$/TiC composites obtained by infiltration techniques. J Phys Condens Matter 16:S2241–S2249
118. Nukami T, Flemings M (1995) In situ synthesis of TiC particulate-reinforced aluminum matrix composites. Metall Mater Trans 26A:1877–1884
119. Yang B, Chen G, Zhang J (2001) Effect of Ti/C additions on the formation of Al$_3$Ti of in situ TiC/Al composites. Mater Des 22:645–650
120. Rajan TPD, Pillai RM, Pai BC (1998) Review: Reinforcement coatings and interfaces in aluminium metal matrix composites. J Mater Sci 33:3491–3503
121. Asthana R (1998) Reinforced cast metal part II evolution of the interface. J Mater Sci 33 (8):1959–1980
122. Leon CA, Lopez VH, Bedolla E, Drew RAL (2002) Wettability of TiC by commercial aluminum alloys. J Mater Sci 37:3509–3514
123. Lumley RN, Sercombe TB, Schaffer GB (1999) Surface oxide and the role of magnesium during the sintering of aluminum. Metall Mater Trans 30A:457–463
124. Orkasov TA, Ponezhev MK, Sozaev VA, Shidov KT (1996) An investigation of the temperature dependence of the surface tension of aluminum alloys. High Temp 34:490–492

Chapter 3
Fabrication Processes for Metal Matrix Composites

3.1 Liquid-State Processes

These processes are the most used today in the manufacture of compounds because they are generally considerably less expensive than solid-state processing and easier to manufacture compared to those of solid-state techniques. However, although they have these advantages, there are some systems that exhibit chemical reactions between the matrix and the reinforcement which sometimes is detrimental for the mechanical properties of composites. Liquid-phase manufacturing is characterized by close interfacial contact and therefore a strong bonding between reinforcement and matrix but can lead to undesirable matrix/reinforcement interfacial reactions producing a brittle interfacial layers. The processes in liquid state can be classified mainly in two types, (1) infiltration and (2) dispersion, as shown in Fig. 3.1. Infiltration processes, in turn, may be assisted by pressure (mechanical or through gas or vacuum) or without pressure (pressureless).

3.1.1 Infiltration Process

Molten metal infiltration in ceramic preforms is one of the most used methods in the manufacture of metal matrix composites. This process involves the infiltration of a molten metal through the channels of a porous ceramic preform to fill the pores and produce the composite. The infiltration of the liquid metal into the matrix can be carried out spontaneously or assisted by an external force (when there is no good wettability between reinforcement and matrix), which may be mechanical produced by an inert gas or by vacuum, mechanical vibration, centrifugal forces, or electromagnetic forces. Some of the advantages of using this process are to infiltrate preforms that have the dimensions and forms close to the piece required and to produce complex shapes with low residual porosity at relatively low cost.

© Springer Nature Switzerland AG 2018
A. Contreras Cuevas et al., *Metal Matrix Composites*,
https://doi.org/10.1007/978-3-319-91854-9_3

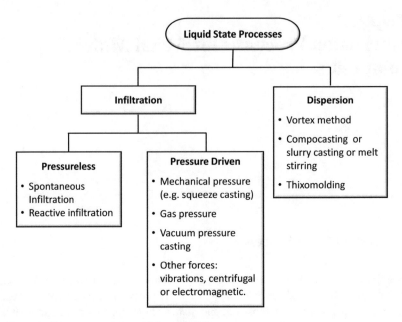

Fig. 3.1 Liquid-state processes to fabricate MMC

The key parameters in infiltration processes assisted by an external force are the initial composition and temperature of the molten metal; the initial composition, volume fraction, temperature, and morphology of the reinforcement; and the nature and magnitude of the external force applied to the metal to overcome the capillary and fluid-drag forces, if there are any [1].

3.1.1.1 Pressureless Infiltration

The infiltration process without external pressure can be divided into spontaneous infiltration and reactive infiltration, depending on the wettability between the reinforcement and the molten metal.

3.1.1.1.1 Spontaneous Infiltration

If there is good wettability between reinforcement and matrix, spontaneous infiltration occurs. If there is no good wettability between reinforcement and matrix, the reinforcement may be coated with some element or compound to improve the wettability and hence the infiltration of the liquid metal into the ceramic matrix. Among the most common reinforcements used are titanium carbide (TiC) and silicon carbide (SiC) infiltrated by aluminum, magnesium, and its alloys. Aluminum alloys are a potential material for most metal-ceramic composites because of its

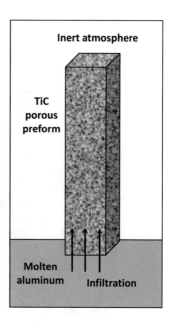

Fig. 3.2 Pressureless infiltration process (Al/TiC system)

combination of low density and high thermal conductivity. Nevertheless, it has the disadvantage of low mechanical strength and a high thermal expansion coefficient. The addition of TiC or SiC improves both its mechanical properties and its resistance to high temperature. On the other hand, magnesium and its alloys have a concern in the automotive industry due to its light weight. The magnesium alloys have a specific resistance compared to the conventional alloys, and some alloys, such as AZ91E, have an excellent castability, good machinability, and good corrosion resistance.

Some examples of this technique fabricated by the authors are the compounds made from sintered titanium carbide preforms infiltrated with aluminum alloys, such as alloy 2024, 6061, and 7075 [2], or magnesium [3] or AlN preforms infiltrated with magnesium [4] or magnesium alloy [5]. The use of these aluminum alloys has some advantages since such alloying elements reduce the contact angle between solid-liquid phases; hence, this would lead to a greater infiltration rate. Furthermore, these alloys can be heat-treated, and therefore the mechanical properties of the composite material may be further improved [6, 7].

The manufacturing process of the TiC/aluminum alloy system is shown in Fig. 3.2. In the infiltration process, sintering and infiltration temperature play a very important role in the kinetic of infiltration. Therefore, the preforms were sintered at 1250, 1350, and 1450 °C during 1 h for obtaining different densifications in order to study its effect on the infiltration process and on the properties of the composites. The porous preforms obtained were infiltrated at temperatures ranging from 900 to 1200 °C using a thermogravimetric analyzer (TGA) for obtaining the infiltration profiles monitoring continuously the weight change of preforms partially immersed in molten aluminum alloy in an inert atmosphere.

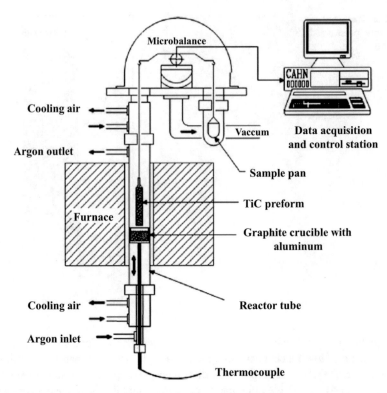

Fig. 3.3 Thermogravimetric analyzer (TGA) setup used for infiltration

Figure 3.3 shows the experimental setup used. The porous preform was suspended from TGA arm by means of a wire to place it in the hot zone of the furnace. A crucible containing solid pieces of aluminum alloy was placed just below the preform. When the aluminum melted and the infiltration temperature was reached, the crucible was raised using the supporting thermocouple rod to partially immerse the end of the preform into the bath aluminum. The mechanical properties of the composites produced depend on the ceramic content and the manufacturing conditions. However, these composites have good physical and mechanical properties due to the high volume fraction of the reinforcement phase.

Pressureless infiltration technique is an attractive process to manufacture metal-ceramic composites because it allows the fabrication of materials with a high ceramic content without the use of an external force. However, for spontaneous infiltration, the molten metal must wet the ceramic; the contact angle between the surfaces of both materials should be <90°, so the capillarity exists; and the metal can be driven into the porous preform. If the contact angle is greater than 90°, the molten metal will not wet the ceramic material, and therefore no infiltration into the porous preform will occur, and therefore the application of an external force or the addition of alloying elements that improve such wettability will be required.

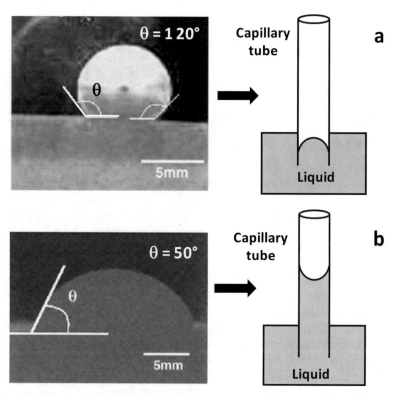

Fig. 3.4 Wettability phenomenon: (**a**) for liquid-solid contact when there is no wettability, no infiltration will occur; (**b**) for liquid-solid contact when wettability exists, infiltration will occur

The contact angle θ is the angle that the liquid forms with respect to the contact surface with the solid and is determined by the resultant of the adhesive and cohesive forces. Figure 3.4 shows both cases, when the molten metal wets the ceramic metal and when wettability does not exist. This angle can be measured commonly by an experimental method known as sessile drop (figures on the left side of Fig. 3.4). The data obtained from this method estimate the liquid-solid interaction, although theoretical model exists for computing the θ values. In the chapter labeled "Wettability," the phenomenon of wettability, including these models, as well as the sessile drop tests is described more deeply.

As mentioned above, it is very important that there is a good wettability between reinforcement and matrix to obtain a strong link between them; however, during the manufacturing of composite in liquid state at temperature relatively high, the alloy is in contact with reinforcement, which in some cases, depending on the conditions of temperature and time, will generate interfacial reactions producing brittle products [6–8], which will produce a weak bond between both materials (matrix and reinforcement).

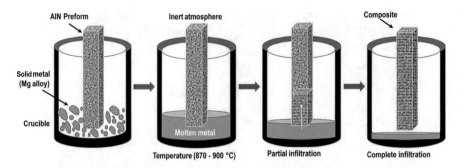

Fig. 3.5 Schematic representation of infiltration process of AlN/Mg system

Regarding the system AlN/magnesium or magnesium alloy, this composites offer good characteristics for its use as a functional material, such as in electronic packaging industry. AlN polycrystalline has a thermal conductivity of 80–200 Kw/ m K and a coefficient of thermal expansion of $4.4 \times 10^{-6}/°C$ (close value of SiC $3.2 \times 10^{-6}/°C$). These two properties made AlN an excellent material for electric circuits of high density compared with other ceramic substrates that have low coefficients of thermal conductivity and high thermal expansion. Therefore, the combination of the AlN properties with the magnesium and its alloys gives an origin to an attractive composite for electronic and structural applications.

AlN preforms were sintered in a temperature range of 1450–1500 °C during 1 h in a nitrogen atmosphere with porosities around 48–51 vol.% and infiltrated at a temperature range of 870–900 °C. The sintered preforms were placed in contact with small pieces of magnesium/magnesium alloy in a graphite crucible inside a horizontal tubular furnace at the infiltration temperature for 10 min in inert atmosphere, and once melted the alloy infiltrated in the preform, obtaining the composite material. Figure 3.5 shows a schematic representation of the AlN/Mg fabrication by pressureless infiltration process. As for the TiC system and aluminum or magnesium alloys, this system also has interfacial reactions [4, 5, 7].

3.1.1.1.2   Reactive Infiltration

This process has been used for the manufacture of NiAl and fiber-reinforced NiAl composites [9]. Nickel wire preforms, in which tungsten fibers may be added, are infiltrated by molten aluminum. The aluminum reacts exothermically with the nickel to synthesize nickel aluminide. A simplified scheme of the process is shown in Fig. 3.6. The continuous fibers can be tungsten, alumina, or molybdenum, which have excellent thermodynamic stability with NiAl. Through this manufacturing technique, pore-free NiAl-W composites with good creep properties are obtained. Composites exhibited compressive creep properties at 715 °C and 1025 °C. At 715 °C, the NiAl-W exhibited secondary creep with little primary and tertiary creep, while at 1025 °C the composites show all three stages.

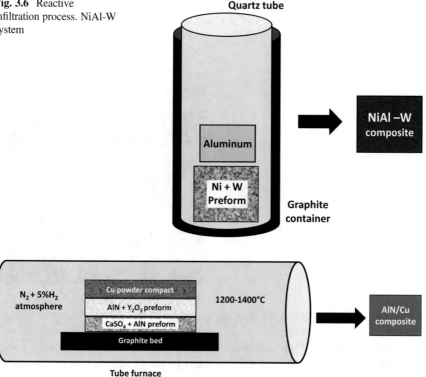

Fig. 3.6 Reactive infiltration process. NiAl-W system

Fig. 3.7 Reactive infiltration process used to fabricate the AlN/Cu composite

Another example of the reactive infiltration process is the AlN/Cu composite [10]. This process is used because of the very low wettability between AlN and the molten copper. AlN preforms with the addition of Y_2O_3 y $CaSO_4$ were infiltrated with Cu powder compact under N_2-5% H_2 atmosphere at temperature range of 1200–1400 °C using a tube furnace during 1 h. Due to the presence of some Cu compounds, infiltration of copper melt into a powder compact of aluminum nitride was carried out. Figure 3.7 shows the setup used for the infiltration process. An AlN/Cu composite with a thermal conductivity of 100 W/m K was obtained.

3.1.1.2 Pressure-Driven Infiltration

When there is no good wettability between the reinforcement and the matrix, the application of an external force is necessary to induce infiltration into the porous preform. This external force can be of the mechanical type, induced by gas or vacuum, by vibrations, by centrifugal forces, or by electromagnetic forces.

Fig. 3.8 Pressure-driven
infiltration process

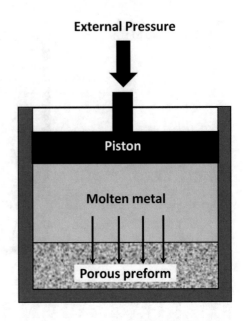

3.1.1.2.1 Squeeze Casting or Pressure Casting

The most common method is known as squeeze casting or pressure casting, which
consists of the application of an external force (a low controlled pressure) by means
of a plunger to force the molten metal into a porous ceramic preform without
damaging it. When the infiltration is complete, a high pressure is applied with the
aim of eliminating the shrinkage and porosity that could remain when the molten
metal contracts due to solidification, thereby forming the composite. Figure 3.8
shows a scheme of this manufacturing process. The reinforcement materials included
carbides, oxides, nitrides, carbon, and graphite, which can be continuous and
discontinuous fibers or particulate. The metallic materials used can be aluminum,
copper, magnesium, and silver. The volume fraction of the reinforcement varies
from 10 to 70 depending on the specific application. The infiltration can be assisted
by vacuum. With this process a good wettability, a complete consolidation, and a
lower or absence of porosity are obtained producing a MMC with excellent mechanical properties. However, expensive molds and large capacity presses are needed.

3.1.1.2.2 Gas Pressure Infiltration or Pressure Infiltration Casting (PIC)

PIC is similar to squeeze casting except that gas, instead of mechanical pressure, is
used to support the consolidation. In this process, the molten metal infiltrates the
porous preform with an inert gas applied from the outside [11]. The melting and the
infiltration of the metal are carried out in a suitable pressure vessel. Usually, the
argon is used as the inert gas at pressures of 150–1500 psi. There are two methods to

Fig. 3.9 Inert gas pressure infiltration process

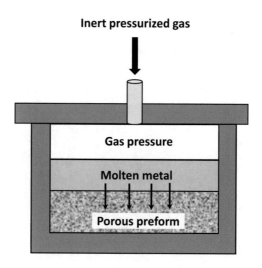

carry out this infiltration process. The first method consists of dipping the heat up preform into the molten metal and then the gas pressure applied to the molten metal surface, resulting in infiltration. Figure 3.9 shows this method. The infiltration pressure will depend on the wettability of the preforms and the volume percentage of the reinforcement, among other factors. In the second method, the molten metal is pressed to the preform by the applied gas pressure using a vertical pipe, and therefore the infiltration occurs. With this process completely dense composites are obtained, with the presence of pores. More reactive materials can be used by this process because of relatively short reaction time.

3.1.1.2.3 Vacuum Pressure Casting

This process is carried out creating a negative pressure differential around the reinforcement to provoke that the liquid moves through preform interstices over-coming the forces of surface tension, viscous drag, and gravity, such as is shown in Fig. 3.10. This process has been used to infiltrate SiC preforms with aluminum (AVCO Specialty Materials) and alumina preforms with aluminum-lithium alloys (DuPont). Another example of the vacuum infiltration process is the infiltration of magnesium into Al_2O_3 or SiC preforms. Commonly, the vacuum infiltration process is used in conjunction with wettability enhancement methods [12]. In the case of lithium with alumina, the lithium can reduce alumina; however this reduction can degrade it; therefore the amount of lithium has to be strictly controlled. In the case of SiC or Al_2O_3 preforms infiltrated with magnesium, the molten metal or its vapor reacts with the air above the preforms to form MgO; therefore the vacuum created drive the infiltration.

Fig. 3.10 Schematic
representation of the
vacuum infiltration process

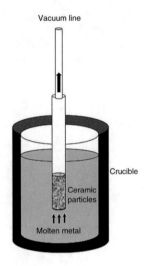

3.1.1.2.4   Other Forces

Other forces used to promote the infiltration of molten metal inside of a porous preform are the vibrations used for infiltrating Al-Si alloys in alumina preforms at frequencies of 3 kHz, centrifugal forces used to produce tubular reinforced metal, and electromagnetic forces [13, 14].

3.1.2   Dispersion Processes

These fabrication processes consist of the addition of solid particles as reinforcement to a liquid light metal matrix as shown in Fig. 3.11. Due to the poor wetting that exists between the reinforcement and the metal, the application of a mechanical force is required to blend both materials, usually through stirring. Dispersion processes are currently the most cost-effective to produce MMCs in large quantities of composite, which can be further processed via extrusion or casting. The simplest method is the vortex method, which consists of strong stirring of the molten melt and the addition of particulate in the vortex. This process has been used to manufacture composite of aluminum and silicon carbide under vacuum and reduced vortex, limiting the incorporation of impurities, oxides, or gases. The particles may be added to the molten metal for further integration by mixing by a rotor or may be injected by a gas beneath the surface of the molten metal. The reinforcement can be particulates, whiskers, or short fibers.

Particles can also be added when the metal is between solidus and liquidus temperature. This manufacturing process is known as compocasting, slurry casting, or melt stirring. Melt metal is vigorously stirred with the solid reinforcement particles to produce a slurry of fine, spheroidal solids floating in the liquid. Stirring

Fig. 3.11 Schematic illustration of a dispersion process

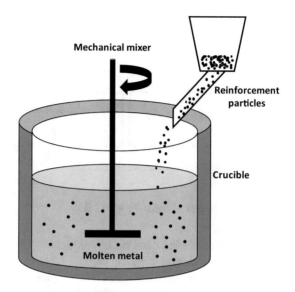

continues during the solidification until the metal itself becomes semisolid and traps the reinforcement particles in a homogeneous dispersion. The slurry produced can be cast by gravity casting, centrifugal casting, die casting, or squeeze casting. A homogeneous reinforcement distribution is obtained through the compocasting process, and a good bonding is achieved between particles and metal matrix. This process is one of the inexpensive methods of fabricating MMCs with discontinuous fibers.

Another process to add particles to the metal in semisolid phase is the process known as thixomolding [15–17]. In this process, the metal pellets and particles are extruded through an injection-molding device.

3.2   Solid-State Processes

These processes have some advantages with respect to liquid processes, such as minimal segregation and reactions of fragile products between reinforcement and matrix. The solid-state process is more suited for reactive systems. Therefore, these fabrication processes are generally used to achieve the highest mechanical properties, particularly in discontinuous MMCs, but are relatively expensive. These processes include powder metallurgy (PM), mechanical alloying (MA) [18–22], diffusion bonding or roll bonding [23–31], and coextrusion and high-rate consolidation [32], which included mechanical alloying.

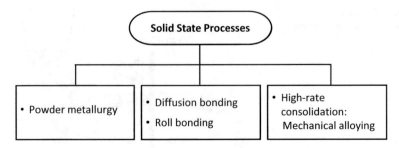

Fig. 3.12 Classification of solid-state processes

3.2.1   Powder Metallurgy (PM)

Figure 3.12 shows the different techniques used in the solid-state process. In solid-state process, the most common method is the powder metallurgy. PM is the most common process for fabricating metal-ceramic and metal-metal composites and more economical than many other fabrication processes. PM is usually used for high melting point matrices, and because of segregation effects and brittle reaction, products that occur in liquid-state processes can be minimized, as well as high residual stresses from solidification shrinkage. This process allows obtaining discontinuously reinforced MMCs with higher mechanical properties to those obtained by some liquid-processes.

Generally, this technology consists of blending powders of the metallic matrix with the reinforcement completely. The blending powders usually are produced by gas-atomized matrix alloy, and the reinforcement can be particulates (powders), platelets, or whiskers. Blending can be performed dry or in liquid suspension. Achieving a homogeneous mixture during blending is a critical issue because the discontinuous reinforcement tends to persist as agglomerates with interstitial spaces too small for penetration of metal powders. These agglomerates are formed due to the large size difference between metal powder and reinforcement, the particle size of the metal powder is commonly from 25 to 30 μm, and the ceramic particulates are often much smaller, from 1 to 5 μm [20]. The obtained mixture is then fed into a mold of the desired shape, and cold pressing is then utilized to compact the homogeneous mixture, obtaining a green compact (preform), which is approximately from 75 to 80% dense. An organic binder can be added to assist to retain the shape of the preform, though these binders often have residual contamination that causes deterioration of the mechanical properties; therefore the binder formulations should be carefully chosen. The preform, which has an open interconnected pore structure, is canned in a sealed container and thoroughly outgassed to low temperature (400–500 °C) to remove volatile contaminants (lubricants and mixing and blending additives), adsorbed gases, and trapped air and vapor water. This step is carried out in long times, from 10 to 30 h. Afterward, the preform is heated at a higher temperature, which is below the melting point but high enough to drive significant solid-state diffusion (process named sintering) while being pressed to

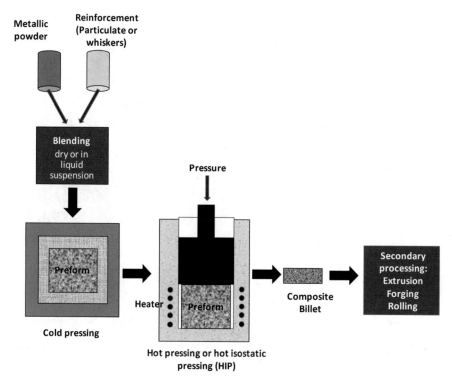

Fig. 3.13 Powder metallurgy process to fabricate MMC

consolidate the metal powder and bond it with the reinforcement. It is recommended that the processing temperature must be above the solidus temperature of the metal to achieve good wetting with the reinforcement. After the bending, the mixture can be also be consolidated directly by vacuum hot uniaxial pressing or hot isostatic pressing (HIP) at a temperature above melt temperature to obtain a fully dense composite, but the reinforcement can be degraded by the applied pressure. The obtained composite billet is then available for a subsequent secondary processing. Secondary fabrication can involve all normally applied metal working processes, such as forging, extrusion, rolling, or drawing for obtaining a final product. Figure 3.13 shows the PM process sequence employed to produce MMC.

Powder metallurgy is used to obtain high volume fractions of particulate, resulting in better properties, and offers near isotropic properties to the composite body; however the powders are expensive, and the blending step is a time-consuming, expensive, and potentially dangerous operation. Moreover, the composite obtained, after the hot pressure is applied, requires a secondary processing for obtaining a final product.

PM processing has been used to produce aluminum MMCs with SiC particulates and whiskers, and Al_2O_3 particles and Si_3N_4 whiskers have also been employed [12, 21, 22]. Conventional aluminum wrought alloys such as 2xxx (e.g., 2124), 6xxx

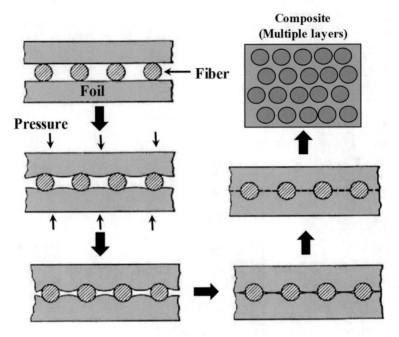

Fig. 3.14 Schematic illustration of the sequence of foil-fiber consolidation process (a diffusion bonding process) and cross-sectional view of the composite

(e.g., 6061 and 6092), and 7xxx (e.g., 7091 and 7090) series can be produced. Other alloys employed include the experimental high-temperature Al-Ce-Fe and the light-weight Al-Li-Cu-Mg [33].

3.2.2 Diffusion Bonding

Diffusion bonding is a solid-state creep deformation process which is used for consolidating thin sheets of the metal matrix material (named foil) with a reinforcement layer (fibers), alternating the foils and fibers layers, producing a sandwich structure (Fig. 3.14). This process is also known as foil-fibber-foil process. The fiber layer is placed between the two foil layers. Several sandwich structures can be done. The metal-fiber sandwich is then compressed (the foils assist to retain the fibers in their place) and heated above the melting point of the metal. This heating with compression causes the metal foils to be melted and diffused toward the fibers, thus wetting the fibers and distributing the metal all over the structure. Diffusion bonding is performed either by loading a vacuum pack in a hot press or by a hot isostatic pressing operation degassing of the pack prior to consolidation.

Diffusion process employed lower temperatures than hot-pressing and reduces fiber/matrix interactions but has some disadvantages, like the applied pressure that it can cause degradation of the fibers; therefore care should be taken to maintain low pressure during consolidation. Furthermore, obtaining complete flow in the space between the fiber midplane and the foil segments on either side is very difficult. The high processing temperatures can cause interfacial reactions between fiber and matrix, producing degradation in the interface decreasing their ability to support load. Coating can be used to reduce the thermal expansion mismatch between the fiber and the matrix, which can commonly cause residual stresses, resulting in a matrix cracking during the cooling from the diffusion bonding temperature. Coatings of Ta, Nb, or other metals have been used on the interface of titanium aluminide (matrix) and SiC fibers. Other disadvantages of the foil-fiber-foil process are the poor fiber distribution with some fibers touching, which cause a detrimental effect on mechanical properties, especially fatigue crack nucleation, and are not capable of net-shape parts except simple shapes. The diffusion process is complicated by the fact that the surface of the foil or metal coating tends to be oxidized, and the oxide makes the bonding more difficult; therefore a vacuum is commonly required for diffusion bonding.

Some composites produced by diffusion bonding process are the aluminum or titanium alloys reinforced with SiC, B, or other fibers. Due to their high properties at elevated temperatures (strength and stiffness), these composites have applications in gas turbine engines.

Roll bonding and coextrusion have also been used for solid-state bonding of metallic foils and fibers and ceramic particulates in metallic matrix [23–31]. Powder blends packed and evacuated in a container can be subjected to these consolidation methods. Laminated composites are ideally produced by high-temperature roll-bonding operations starting from either foils of individual metals or from alloys. During roll bonding, both surface deformation and diffusion actively cause rough-ness deformation and interdiffusion to produce strong interfaces. The multilayer of Cu-Nb composites has been produced by roll bonding, and the rods of Cu and Ni (or W, Nb) can also be coextruded so that both metallic phases deform and lead to the fine microfibers of the harder phase [26, 27].

Fabrication of complex-shaped components is commonly achieved by diffusion bonding of monolayer composite tapes. The tapes may be prepared by a number of methods, but the most commonly used is filament winding, where the matrix is incorporated in a sandwich construction by laying up thin metal sheets between filament rows or by the arc spraying technique. The obtained monotape commonly has some porosity degree, but this can be eliminated during diffusion bonding or during HIP. The monotapes can be stacked in layers, and using the diffusion bonding method, the composite can be obtained. Some examples of composites produced by this process are the W-fiber-reinforced Ni-based alloy monotapes, which have been used for the fabrication of lightweight, hollow turbine blades for a high-performance engine. The process involves placing monotapes around a bent steel core and diffusion bonding several monotape layers. When the steel core is eliminated by acid, near-net shape blades are produced. This method is being used to fiber-reinforced titanium matrix composites to improve fiber distribution and improve the mechanical properties of the composite material.

3.2.3   High-Rate Consolidation: Mechanical Alloying (MA)

This consolidation process of powder blends is most suitable for rapid solidification (RS) metals and hard-to-deform metals. Frictional heating at the powder-particle interface causes local melting and consolidation, and rapid heat extraction by the cooler particle interior causes RS. Thus, an RS microstructure can be better preserved by this method. High-rate consolidation leads to strengthening of alloys (due to high dislocation density) but reduced ductility [32].

Mechanical alloying is a truly solid-state process that can be applied to particulate composites, intermetallic compound, and alloys with microstructures that are not possible to obtain by standard metallurgical practices such as casting and forging. This process consists in simultaneous and repeated sequences of extensive plastic deformation, cold welding, fracturing, and re-welding of a powder mixture as fresh internal surfaces are exposed. The continuous fragmentation leads to thorough mixing of the constituents, and subsequent processes such as hot-pressing and extrusion are used to consolidate and/or synthesize the alloy or composite.

MA process requires employment of high-energy mills like the attrition, constituted by a static drum and a rotor (Fig. 3.15). The rotor is constituted for a drive shaft on which a series of arms or rotational impellers are mounted. The motor moves the rotor, and the impellers shake the steel balls and the powder inside the drum. Using this equipment can reach mill speeds ten times bigger than those of a conventional mill. Other equipment used in this process is the vibratory and planetary mills, large ball mills, and high-speed blenders and shakers (e.g., high-energy Spex).

During the milling, the powders suffer a series of processes that are initiated when the balls catch some of their particles among them, smoothing them, breaking them, and joining them to other particles. Every time those collisions take place among

Fig. 3.15 Attrition ball mill used for mechanical alloying

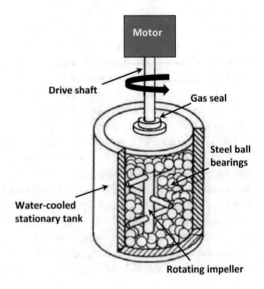

balls, the contained particles are deformed, creating in this way new surfaces. These new surfaces have great tendency to be oxidized; for this reason the mill process should be carried out in the absence of air, using vacuum or in inert gas atmosphere. In the first stages of the mill process, the particles of metallic powder are soft and ductile, so being crushed by the balls, they have a tendency to be welded forming powder particles with structure of layers. In consequence, during this stage, the size of the particles usually increases. As the process advances, due to the great quantity of energy introduced in the form of plastic deformation, the particles become hard and brittle. In this sense, the biggest particles have more probability to incorporate cracks and to break them when they are hit by the steel balls. This welding process in cold and fracture occurs repeatedly during the milling. Therefore the structure of the particles of powders is continually refined and homogenized. Finally, the tendencies to be welded and to break are stabilized, achieving a dynamic balance inside a narrow margin.

Currently, a great deal of interest has developed the use of mechanical alloying as a way of dispersing a ceramic phase more uniformly within the metallic matrix in the blend. By repeated impact of the powder particles and their fragmentation, a fresh metallic surface is constantly exposed, within which the ceramic particles become embedded. Particle distribution has been shown to be greatly improved by this method. NiAl and Ni_3Al composites containing Al_2O_3, Y_2O_3, ThO_2, and AlN particulates have been fabricated using this process. The obtained powder can be canned and extruded to create bulk composite specimens. This process has also been used to produce ultrafine aluminum nitride (AlN) dispersion using milling NiAl as matrix with finely divided Y_2O_3 in liquid nitrogen, a process named as cryomilling or reaction milling, due to AlN reinforcement forms through a chemical reaction of aluminum with nitrogen during milling [34].

MA has the same advantages as rapid solidification (RS), such as extension of solubility limits, production of novel structures, and refinement of the microstructure (down to the nanostructure range), and it enables the production of a dispersion of second-phase particles. However, a challenge in MA is the need to separate the attrited powders from the unattrited powders to obtain a homogeneous material. Another issue of the MA is the difficulty that arises from contamination introduced from the vessel, balls, hammers, or surfactants used in the pulverizing mill and from the chemical reaction that occurs with the atmosphere within the mill. This can happen, especially, when milling certain powders of very hard materials, due to the waste and later incorporation to the mill of small portions of material of the balls and walls of the mill. With proper care, it is possible to keep such problems to a minimum. It is possible to carry out the mill with vessels and balls of a harder material to the one that is required to alloy. Furthermore, container materials must be ductile and nonreactive to the powder mixtures [34]. In most cases, however, contamination leads to the undesirable presence of oxides, carbides, Fe, or W in the product, but this problem can be solved. For example, an extremely fine grain size (<1 μm) was stabilized by a fine dispersion of ceramic particles obtaining a high degree of superplastic ductility in 6061 Al/SiC and 6061Al/Si_3N_4 particulate composites prepared by mechanical alloying [21, 22].

The other great problem of MA is that it is a process relatively expensive which limits its industrial application. However, the cost, in certain cases, is not so important, if they are kept in mind that materials produced by MA have better properties to those produced by other fabrication methods. The low productiveness is due to the milling times which are usually long.

3.3   Gas- and Vapor-Phase Processes

The gas- and vapor-phase processes can be classified mainly into two: the spray processes and the vapor deposition processes, such as is shown in Fig. 3.16.

3.3.1   Spray Deposition

These processes consist of the fragmentation (by means of a high-speed cold inert gas jet, commonly argon or nitrogen) of a stream of molten metal in fine droplets (300 μm or less) that are sprayed together with the reinforcement (if the reinforcement is a particulate) and collected on a substrate or mold where the semisolid metal droplets recombine and solidify to produce the composite material, such as is shown in Fig. 3.17 [33, 35]. When the droplets impact the substrate or mold at very high velocities in molten or partially solidified state, they deform (flatten) into splats and weld together to form the composite, as it is shown in Fig. 3.18, which shows the

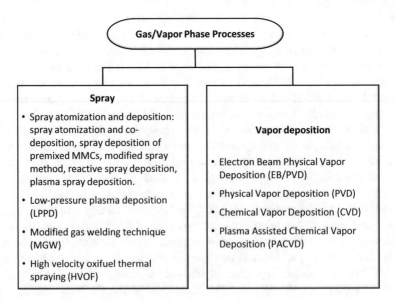

Fig. 3.16 Gas- or vapor-phase processes

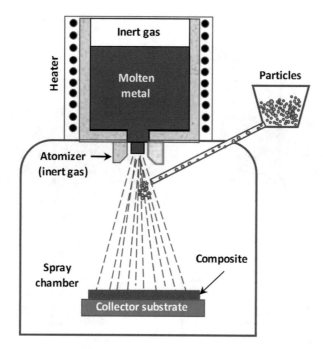

Fig. 3.17   Spray deposition process

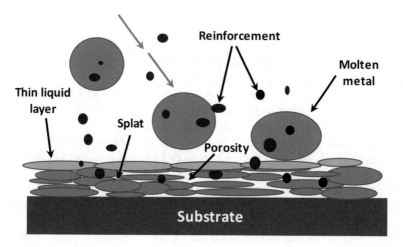

Fig. 3.18   Sketch showing co-deposition of a metallic matrix and ceramic particulates via a thermal spray process

basic concept of spraying and particle integration in the solidified matrix [34]. Although the high energy of impact assists in powder consolidation and densification, a high degree of porosity commonly remains in the as-deposited material; therefore a subsequently forged or other secondary process to form a

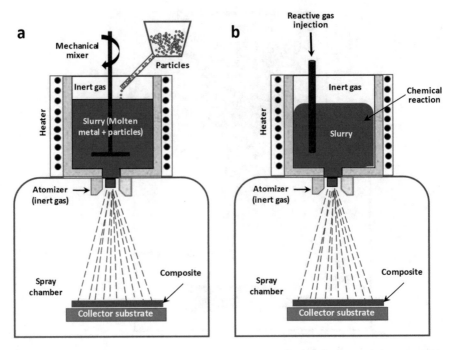

Fig. 3.19 Spray deposition synthesis of premixed MMCs, (**a**) slurry is formed by molten metal and particles, and (**b**) slurry is formed by molten metal and reinforcement (produced by chemical reaction between molten metal and gas)

fully dense product is required. The porosity levels depend on the thermal conditions, impact velocity, and spray density or mass flux.

The spray process is a hybrid rapid solidification method because the metal experiences a rapid transition from the liquidus to the solidus, followed by slower cooling from the solidus to room temperature [12, 20, 34].

Another variation of the spray process is the reactive spray-forming by gas-liquid reactions, in which the spray nozzle is extended to include a reaction zone where reactive gases are fed into the liquid stream [36, 37]. The gases react with the molten particles (droplets) to form the reinforcement, as it is shown in Fig. 3.19. Carbide, oxide, silicide, and nitride reinforcement in metal matrices have been produced in this manner [34].

The reinforcement can also be applied on the substrate, and the molten metal may be sprayed onto it. The production of MMC ingot by spray deposition can be accomplished by introducing reinforcement into the metal spray, driven to co-deposition with the atomized molten metal onto the substrate. Careful control of the atomizing and reinforcement feeding conditions are required to guarantee that a uniform distribution of particulate is produced.

The as-sprayed material has a density within 90–98% and shows a uniform distribution of fine equiaxial grains and no prior particle boundaries or discernible macroscopic segregation. Mechanical properties are normally isotropic and meet or exceed those of ingot-processed alloys. High rates of metal deposition in the range 0.2–2.0 kg/s are obtained with this process [12, 20].

Spray deposition was implemented commercially in the late 1970s and during the 1980s by Osprey Ltd as a method of producing bulk material by atomizing a molten stream of metal with jets of cold gas. However, this process has been tailored to particulate MMC production by injection of ceramic powder into the spray. The Osprey method has four steps: (1) melting and dispensing, (2) gas atomization, (3) deposition, and (4) collector manipulation. In the production of the melt, an induction heating is used, which flows into a gas atomizer. Melting and provision are performed in a vacuum chamber. The atomized stream of metal is collected on a substrate placed in the line of flight. Overspray is separated by a cyclone and collected [20].

Several composite spray-based processes exist, which vary in the way of spraying the molten metal and in the (often proprietary) method the reinforcement is mixed with the molten metal. Most such processes are covered by their patents or licenses. Among these processes are the spray atomization and deposition, spray atomization and co-deposition, spray deposition processing of premixed MMCs, low-pressure plasma deposition (LPPD), modified gas welding technique (MGW), and high-velocity oxyfuel thermal spraying (HVOF) [33]. Most of these processes utilize gases to atomize the molten metal into fine droplets (usually up to 300 μm diameter). The spray atomization and co-deposition is the first and most frequently employed for the fabrication of composite materials. The expression co-deposition is used to describe simultaneous deposition of both the metal matrix and reinforcing phase. In the process of spray deposition of premixed MMCs, the starting materials for atomization and deposition are composites, which the reinforcement is added by using one of the two methods shown in Fig. 3.19. The LPPD process was developed for the net-shape processing by plasma spraying in a reduced pressure environment using pre-alloyed powders, different to the process of conventional atomization, which is performed at atmospheric pressure. In the MGW process, the gas metal arc (GMA) welding torch is modified where wire feedstock is melted and combined with the reinforcement entrained in an inert gas, and then the mixture is deposited in a substrate or mold solidifying into a composite structure. In the HVOF process, the powder flow is electronically controlled, and feed rates are monitored automatically [33].

The parameters that govern the spray processing are the initial temperature, size distribution, and velocity of the molten metal droplets; the velocity, temperature, and feed rate of the reinforcement (when this is injected simultaneously); and the position, nature, and temperature of the substrate collecting the material [1, 20]. In order to use the spray deposition process it is necessary to understand and control the effects of several independent process parameters, i.e., melt superheat, metal flow rate, gas pressure, spray motion (spray scanning frequency and angle), spray height (distance between the gas nozzles and the substrate), and substrate motion (substrate rotation speed, withdrawal rate, and tilt angle) [12].

The reinforcement can be added within the droplet stream or between the liquid stream and the atomizing gas (e.g., the Osprey process) or by continuous feeding of cold metal into a zone of rapid heat injection (e.g., thermal spray processes) [38].

The spray processes have been used to manufacture composites of aluminum alloys (aluminum-silicon casting alloys and the 2XXX, 6XXX, 7XXX, and 8XXX series wrought alloys) reinforced with SiC, Al_2O_3, or graphite particles. Products that have been produced by spray deposition include solid and hollow extrusions, forgings, sheet, and remelted pressure die castings [12, 20]. Arc melting, flame spraying, or a combination of both has been utilized to produce molten metal drops from aluminum wires. For high-temperature materials, plasma torches are used to melt and spray metal powders [39] (e.g., Ni_3Al-reinforced TiB_2 particles, as well as various low- and high-temperature matrices (e.g., Al alloys and Ti_3Al or $MoSi_2$, reinforced with SiC, TiC, or TiB_2 particles)). Composite monotapes of continuous fibers and Ti-based matrices are also prepared by plasma spraying [39, 40], to be further processed in the solid state, for example, by diffusion bonding, as it was mentioned previously in Sect. 3.2.2. These monotapes are produced from continuous fibers that are wrapped around a mandrel with controlled interfiber spacing, and the matrix metal is sprayed onto the fibers. Subsequently, the composites are formed by hot-pressing of composite monotapes. Fiber volume fraction and distribution are controlled by adjusting the fiber spacing and the number of fiber layers. With continuous ceramic fibers, however, the large thermal expansion mismatch between the ceramic fiber and the metal matrix and the thermal shock and mechanical stresses arising from the initial exposure to plasma jet can result in displacement and fracture of fibers, especially because the brittle ceramic fibers are already bent to large curvatures when wrapped on a mandrel [34].

Due to the high solidification rates (in the range of 10^3–10^6 K s^{-1}) of the droplets in the spray deposition processes, little segregation, minimal reinforcement degradation, very low oxide content, matrices with fine grain sizes, and precipitation structure with no significant increase in solute solubility are obtained in the fabricated composites. A strong interfacial bond and little or no interfacial reaction layers are also obtained because liquid metal and reinforcement contact only briefly (no more than a few tens of milliseconds), permitting the production of thermodynamically metastable two-phase materials (e.g., iron particles in aluminum alloys). Even for small particles, chemical interactions between the reinforcement and the molten metal will not alter the composition and properties of the interface, although thin interfacial layers of reaction compounds (e.g., intermetallics) may form by diffusional interactions that may improve the interfacial bond strength and hence the composite properties [34]. Ingots or tubes can be produced by these processes, although they lack in part-shape versatility [1, 20]. The disadvantages of spay deposition process include the following: (a) an inhomogeneous distribution of the reinforcement particles resulting in ceramic-rich layers normal to the growth direction, with significant residual porosity levels of about 5% (which requires subsequent processing to achieve a full consolidation); (b) the process is more expensive than casting or infiltration processes, due to the longer processing times; (c) the high cost of the gases used; and (d) the large amounts of waste powder to be collected and disposed [1, 12, 20].

3.3.2   Vapor-Phase Deposition

Electron beam/physical vapor-deposition (EB/PVD) is a vapor deposition technique used for the fabrication of MMCs [41–43]. This evaporation process consists of passing of fibers by a region of a high partial vapor pressure of the melt to be deposited, where condensation takes place to produce a relatively thick coating on the fiber. The vapor is produced by directing a high power of an about 10 kW electron beam onto the end of a solid bar feedstock with characteristic deposition rates of about 5–10 μm/min. If the vapor pressures of the elements in the alloy are relatively close to each other, a single source for electron beam evaporation can be used; if not, multiple source evaporation should be utilized. Alloy composition can be modified, due to differences in evaporation rates between different solutes which are compensated changing the composition of the molten pool formed at the end of the solid bar feedstock until a steady state is reached in which the alloy content of the deposit is the same as that of the feedstock [38]. An example of this kind of PVD process is the titanium matrix composite reinforced with SiC fibers [20].

Another technique that employs vapor-phase deposition is the chemical vapor deposition (CVD) process, which has been used for forming coatings such as TiB_2, TiC, SiC, B_4C, and TiN on carbon fibers, which are used as precursor wires to fabricate MMCs [44]. The TiB_2 coating is particularly attractive because of the exceptionally good wetting between TiB_2 and molten aluminum. Figure 3.20 shows the fabrication of precursor wires of aluminum-impregnated TiB_2-coated carbon fibers. The TiB_2 coating uses $TiCl_4$ and BCl_3 gases reduced by Zn vapor [45].

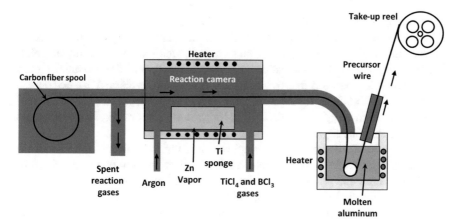

Fig. 3.20 Fabrication by chemical vapor deposition of precursor wire of aluminum-impregnated TiB_2-coated carbon fibers

A low modulus is preferred for the coating with the objective of increasing the interface strength and the transverse strength. Different modulus values (19–285 GPa) of SiC coatings can be obtained by controlling the plasma voltage in plasma-assisted chemical vapor deposition (PACVD). A value of 448 GPa is only obtained by CVD SiC. Aluminum matrix composites reinforced with carbon fibers coated with SiC show an interface strength and a transverse strength which increase with decreasing the modulus of the SiC coating [45].

Another vapor-phase fabrication process is the chemical vapor infiltration (CVI), which involves infiltration of reactant vapors into a porous preform and deposition of the matrix (solid product phase) within the pores [46]. This process is slow and may take up to several hundred hours for completion but have net-shape potential and reduced processing temperatures.

The vapor-phase deposition processes have the following advantages: (a) a wide range of alloy compositions can be used; (b) there is little or no mechanical disturbance of the interfacial region which may be quite significant when the fibers have a diffusion barrier layer or a tailored surface chemistry; (c) very uniform fiber distributions can be produced with fiber contents of up to about 80%; (d) the fiber volume fraction can be perfectly controlled by the thickness of the deposited coatings, and the fiber distribution is always very homogeneous; (e) the time required for diffusion bonding is shorter; and (f) the coated fiber is relatively flexible and can be wounded into complex part shapes [20].

Composite fabrication by these fabrication processes is usually completed by assembling the coated fibers into a bundle or array and consolidating by vacuum hot-pressing of HIP operation.

3.4   In Situ Processes

In situ composites can be fabricated by a broad range of several processes. However, these processes generally can be divided mainly into two types of processes, composites obtained from solidification of a melt, well known as controlled solidification or directional solidification processes [47–50], and those obtained from chemical reaction between phases (melt and solid or gaseous phases) [51–55], as can be observed in Fig. 3.21. Inside these last ones is the exothermic dispersion or self-propagating high temperature (SHS) [56–60]. Some authors consider the in situ processes inside the liquid or solid-state processes.

The term in situ composite was employed initially to fabricate alloys by directional solidification for optics and electronics applications, but due to problems of low growth rates and gradual coarsening of the structure at high temperatures, their use was restricted. In recent times, because of the processing of heat-resistant composites, a new impetus to research on in situ composites has arisen. Schematic examples of in situ processes are given in Figs. 3.22 and 3.23.

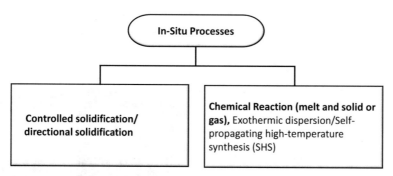

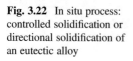

Fig. 3.21   In situ process classification

Fig. 3.22   In situ process:
controlled solidification or
directional solidification of
an eutectic alloy

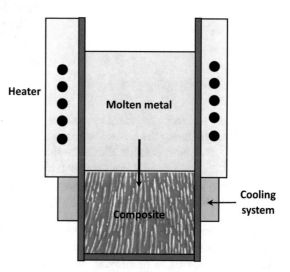

A major advantage of in situ composite materials is that the reinforcing phase is generally homogeneously distributed, and spacing or size of the reinforcement may be adjusted in several cases by the solidification or reaction time. The interfaces are clean and mutually compatible because the constituent phases crystallize in situ rather than combined from separate sources. However, the system selection and the reinforcement orientation are limited, and the process kinetics (in the case of reactions), or the shape of the reinforcing phases, is sometimes difficult to control.

3.4.1   Controlled or Directional Solidification Process

As it was mentioned previously, one of the methods of production of in situ composites is the controlled or directional solidification of two or more phase alloys containing in situ-grown reinforcement (Fig. 3.22) [47–50]. One of the phases

Fig. 3.23 In situ process:
chemical reaction process
(gas-molten metal)

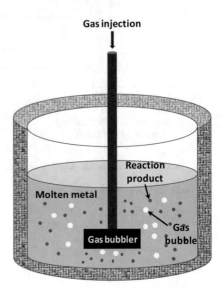

solidifies in the form of fibers or rodlike structures (reinforcement) within the
metallic matrix during the controlled solidification.

Usually, for binary eutectic alloys, the spacing between rods or lamellae depends
on the cooling rate and the volumetric fraction of the reinforcement of the alloy
composition. The reinforcement is aligned in the growth or solidification direction.
Taking into account the surface energy considerations, the rodlike morphology is
favored for low volumetric fraction of one phase. For alloys that are constituted by
three or more phases, the reinforcement spacing decreases as the cooling rate
increases, and a wide array of reinforcement shapes and compositions can be
achieved. Moreover, with alloying additions the reinforcement type can be controlled. For instance, Nb or Ta additions to TiAl/B melt produce the formation of TiB
rods, while the solidification of TiAl/B melt without alloying additions produces
TiB$_2$ equiaxial particles.

The materials produced by directional solidification are intrinsically stable
because they are produced under near-equilibrium conditions and usually incorporate low-energy interface boundaries. Nevertheless, a coarsening occurs when they
are exposed to temperature changes or gradients, and the rods tend to thicken and
shorten, modifying the mechanical properties of the composite. The oxidation and
interfacial or bulk diffusion provoked by a change in the solubility of the phases, and
which happens commonly at grain boundaries of faults in the structure, are the
mechanisms responsible for the coarsening [47–50].

Examples of in situ composites are some eutectic systems, such as the Nb-NbC
system; Ni- or Co-based eutectic super alloys; eutectic systems based on Cr and Ta;
Fe-TiC composites produced from solidification of Fe-Ti-C melts; TiB rods in TiAl
matrix from solidification of melts containing TiAl, Ta, and B, as well as TiC/Ti

composites from mixtures of Ti; and C with Al additions and Ni-Al intermetallics (Ni$_3$Al and NiAl) [47–50].

Directional solidification has been used to develop materials with dual-phase microstructures, which have improved toughness, ductility, and creep strength, such as Ni-based composites, in which one of the aligned phase is sandwiched between a ductile second phases. An example is the NiAl-X composites, where X can be Mo, W, Cr, and Fe.

The improvement in the toughness in in situ composites is due to the inhibition of crack nucleation in the secondary phase, inhibition of crack growth due to plastic bridging by the ductile phase (crack bridging), and blunting of the crack by the ductile phase [34]. Such is the case of the dispersed second phase in directionally solidified NiAl-W and NiAl-Cr alloys, which improve the toughness of these alloys by means of the prevention of the crack propagation.

The directional solidification with small alloying additions (such as Mo, Cr, W, and Nb) is broadly used to improve the creep resistance, room temperature ductility, toughness, and the strength of NiAl. Aligned dual-phase microstructure of Ni-43Al-9.7Cr and Ni-48.3Al-1W alloys has achieved improvements in compressive yield strength and ductility as compared to the single-phase NiAl, with W alloying yielding the greatest strength improvements at room temperature. Ni-based alloys produced by directional solidification show a ductility greater than near-zero ductility of NiAl at room temperature, with Cr producing the largest improvement in fracture strain, followed by W.

The spacing between rods or lamellar, cell size, and the second-phase morphology are controlled by varying the growth speed (cooling rate), such is the case of directional solidification of NiAl and Ni$_3$Al. For example, the directional solidification of Ni$_3$Al at cooling rate of 25 mm/h produced a columnar-grained single-phase Ni$_3$Al with ~60% tensile ductility at room temperature, nevertheless, the same material, but at cooling rate of above 50 mm/h caused lower ductility.

The improvement of the strength to high temperature and creep resistance of in situ composites can be achieved by the addition of ceramic fibers as reinforcement in the growth of the compound. For instance, NiAl composites with Cr or W additions were reinforced with sapphire fibers by various directional solidifications. High interfacial shear strength, high-temperature strength, and room temperature toughness are obtained. Using directional solidification, the growth of large, columnar NiAl grains with sapphire fibers residing in the eutectic colonies are obtained. Cr is known to improve the creep strength and produces a significant amount of strengthening due to fine Cr precipitates; Re, Cr, and Mo produced an aligned phase improving the strength.

Some disadvantages of controlled or directional solidification process are the slow rates of growth, approximately 0.4–2.0 in/h (1–5 cm/h), due to the need to maintain a stable growth front, which requires a large temperature gradient. In addition, there are also limitations in the reinforcement nature and volume fraction and the morphological instabilities associated with thermal gradients, which has caused a decrease in the interest in these types of composites.

3.4.2 Chemical Reaction Processes

The in situ composites can also be obtained by means of the chemical reaction between metal molten and solid with a gas. The governing mechanisms of this kind of processes involve the identification of the possible reactions happening and evaluation of the driving force derived from thermodynamic considerations and reaction kinetics, which depends on temperature and alloy, gas, or solid compositions and concentrations, as well as diffusion mechanisms across reaction or boundary layers [51–55].

Some examples of in situ composites produced by chemical reaction are the Al_2O_3/Al composites obtained from aluminum oxidation and TiC-reinforced Al-Cu alloys, which have been obtained by bubbling methane and argon gas through Al-Cu-Ti melt, such as is shown in Fig. 3.23. This gas injection method can be used to produce reinforced alloys with carbides and nitrides. Other examples are the composites of TiB_2-reinforced aluminum alloys made by the exothermic dispersion (XD) process, in which Ti, B, and Al powders were heated at 800 °C so that they react forming TiB_2, TiB whiskers obtained after laser melting of Ti and ZrB_2 powders, or TiAl matrix obtained after squeeze casting of molten aluminum into TiO_2 powders or short fibers.

One of the processes in situ more commonly employed is the XD process, which uses reactions highly exothermic between two reactants to produce a third compound. XD process is also known as "combustion synthesis" or self-propagating high-temperature synthesis (SHS) [56–60]. It is known by this name because the process uses highly exothermic reactions (which reach temperatures of up to 3000 K) to be self-sustaining and energy efficient. A master alloy with a high volume fraction of reinforcement is mixed and remelted with a base alloy and is placed in a special reactor under argon blanket and then ignited using laser beams or other high-energy sources to initiate the reaction, and the reaction continues by the heat released from the reaction until the feed material has combusted and reacted to produce the desired reinforcement amount. An example of this XD process is the mixture of Al, Ti, and B, which are heated sufficiently until an exothermic reaction occurs among them producing a TiB_2 mixture distributed in a titanium aluminide matrix, achieving superior strength levels superior to 100 ksi (690 MPa) at room temperature.

Some advantages of the in situ processes are the following: (a) due to some of these reactions which are highly exothermic, the fabrication process is fast and self-propagating; (b) the reaction products (reinforcements) are thermodynamically stable, driving to less degradation at elevated temperature; (c) the interface between reinforcement and matrix is clean, resulting in a strong interfacial bonding; and (d) the reinforcements produced are finer in size, and their distribution in the matrix is more homogeneous [20, 38].

Some disadvantages of the in situ processes are the high levels of open porosity obtained in the final product, which are smaller to 50%, and full densification for high-performance applications almost requires secondary processing such as hot consolidation, solid-state consolidation, and infiltration to produce composites with high density. For example, the aluminum liquid phase is formed by the following reaction:

$$3TiO_2 + 3C + (4 + x)Al = 3TiC + 2Al_2O_3 + xAl \qquad (3.1)$$

which infiltrates all open porosities and densities of the porous ceramic phase. However, the formation of a liquid phase during combustion synthesis is beneficial because it increases the contact area between reactants and permits faster diffusion and reaction, such as the case of Ti-C system with aluminum, which increases the surface area and the rates of reaction and mass transfer [34].

In summary, the selection of the manufacturing process depends on several factors, among which are the nature of the matrix and reinforcements; quantity and distribution of the reinforcement and the application preservation of reinforcement strength; minimization of reinforcement damage; promotion of wetting and bonding between the reinforcement and the matrix; flexibility that allows proper support, orientation, and spacing of the reinforcement inside the matrix; type of composite selected; production cost; process efficiency; and the quality of the product.

References

1. Suresh S, Mortensen A, Needleman A (eds) (1993) Fundamental of metal matrix composites. Butterworth-Heinemann, Boston
2. Contreras A, Salazar M, León CA et al (2000) Kinetic study of the infiltration of aluminum alloys into TiC preforms. Mater Manuf Process 15(2):163–182
3. Contreras A, López VH, Bedolla E (2004) Mg/TiC composites manufactured by pressureless melt infiltration. Scr Mater 51(3):249–253
4. León CA, Arroyo Y, Bedolla E et al (2006) Properties of AlN-based magnesium-matrix composites produced by pressureless infiltration. Mater Sci Forum 509:105–110
5. Bedolla E, Lemus-Ruiz J, Contreras A (2012) Synthesis and characterization of Mg-AZ91/AlN composites. Mater Des 38:91–98
6. Albiter A, León CA, Drew RAL et al (2000) Microstructure and heat-treatment response of Al-2024/TiC composites. Mater Sci Eng A 289:109–115
7. Reyes A, Bedolla E, Pérez R et al (2016) Effect of heat treatment on the mechanical and microstructural characterization of Mg-AZ91E/TiC composites. Compos Interfaces:1–17
8. Aguilar EA, León CA, Contreras A et al (2002) Wettability and phase formation in TiC/Al-alloys assemblies. Compos Part A 33:1425–1428
9. Venkatesh TA, Dunand DC (2000) Reactive infiltration processing and secondary compressive creep of NiAl and NiAl-W composites. Metall Mater Trans A 31:781–792
10. Hsu HC, Chou JY, Tuan WH (2016) Preparation of AlN/Cu composites through a reactive infiltration process. J Asian Ceramic Soc 4:201–204
11. Kainer KU (ed) (2006) Metal matrix composites custom-made materials for automotive and aerospace engineering. Wiley-VCH Verlag GmbH & Co. KGaA, Weinheim
12. ASM Handbook (2001) Composites, vol 2. ASM International, Ohio
13. Elahinejad S, Sharifi H, Nasresfahani MR (2018) Vibration effects on the fabrication and the interface of Al–SiC composite produced by the pressureless infiltration method. Surf Rev Lett. https://doi.org/10.1142/S0218625X18500890

14. Nakanishi H, Tsunekawa Y, Mohri N et al (1993) Ultrasonic infiltration in alumina particle/ molten aluminum system. J Jpn Inst Light Met 43(1):14–19
15. Midson P, Kilbert RK, Le Beau SE et al (2004) Guidelines for producing magnesium thixomolded semi-solid components used in structural applications. In: Proceedings of the 8th International Conference on Semi-Solid Processing of Alloys and Composites, September 21–23, 2004, Limassol, Cyprus
16. Montrieux HM, Mertens A, Halleux J et al (2011) Interfacial phenomena in carbon fiber reinforced magnesium alloys processed by squeeze casting and thixomolding. In: European Congress and Exhibition on Advanced Materials and Processes, 12–15 Sep. Montpellier, France, pp 1–25
17. Decker RF, LeBeau SE (2008) Thixomolding. Adv Mater Process 2014:28–29
18. Si YG, You ZY, Zhu JX et al (2016) Microstructure and properties of mechanical alloying particles reinforced aluminum matrix composites prepared by semisolid stirring pouring method. China Foundry 13(3):176–181
19. Ma H, Lu Y, Lu H et al (2017) Fabrication of Ni/SiC composite powder by mechanical alloying and its effects on properties of copper matrix composites. Int J Mater Res 108(3):213–221
20. Campbell FC (2010) Structural composite materials. ASM International, Ohio
21. Fogagnolo JB, Robert MH, Torralba JM (2003) The effects of mechanical alloying on the extrusion process of AA 6061 alloy reinforced with Si_3N_4. J Braz Soc Mech Sci Eng 25 (2):201–206. https://doi.org/10.1590/S1678-58782003000200015
22. Fogagnolo JB, Velasco F, Robert MH et al (2003) Effect of mechanical alloying on the morphology, microstructure and properties of aluminium matrix composite powders. Mater Sci Eng A 342(1–2):131–143
23. Testani C, Ferraro F, Deodati P et al (2011) Comparison between roll diffusion bonding and hot-isostatic pressing production processes of Ti_6Al_4V-SiC_f metal matrix composites. Mater Sci Forum 678:145–154
24. Chaudhari GP, Acoff V (2009) Cold roll bonding of multi-layered bi-metal laminate composites. Compos Sci Technol 69(10):1667–1675
25. Luo JG, Acoff VL (2004) Using cold roll bonding and annealing to process Ti/Al multi-layered composites from elemental foils. Mater Sci Eng A379(1–2):164–172
26. Shabani A, Toroghinejad MR, Shafyei A (2012) Fabrication of Al/Ni/Cu composite by accumulative roll bonding and electroplating processes and investigation of its microstructure and mechanical properties. Mater Sci Eng A558:386–393
27. Hosseini M, Pardis N, Manesh HD et al (2017) Structural characteristics of Cu/Ti bimetal composite produced by accumulative roll-bonding (ARB). Mater Des 113:128–136
28. Motevalli PD, Eghbali B (2015) Microstructure and mechanical properties of Tri-metal Al/Ti/ Mg laminated composite processed by accumulative roll bonding. Mater Sci Eng A 628:135–142
29. Muratoğlu M, Yilmaz O, Aksoy M (2016) Investigation on diffusion bonding characteristics of aluminum metal matrix composites (Al/SiC_p) with pure aluminum for different heat treatments. J Mater Process Technol 178(1–3):211–217
30. Lin H, Luo H, Huang W et al (2016) Diffusion bonding in fabrication of aluminum foam sandwich panels. J Mater Process Technol 230:35–41
31. Zhang XP, Ye L, Mai YW et al (1999) Investigation on diffusion bonding characteristics of SiC particulate reinforced aluminum metal matrix composites (Al/SiC_p-MMC). Compos A Appl Sci Manuf 30(12):1415–1421
32. Raghunathan SK, Persad C, Bourell DL et al (1991) High-energy, high-rate consolidation of tungsten and tungsten-based composite powders. Mater Sci Eng A 131(2):243–253
33. Srivatsan TS, Lewandowski J (2006) Metal matrix composites: types, reinforcement, processing, properties, and applications. In: Soboyejo WO, Srivatsan TS (eds) Advanced structural materials; properties, design optimization, and applications. CRC Press/Taylor & Francis Group LLC, Boca Raton, pp 275–357

34. Asthana R, Kumar A, Dahotre NB (eds) (2005) Materials processing and manufacturing science. Elsevier Science & Technology Books, London
35. Haghshenas M (2015) Metal-matrix composites. Elsevier, https://doi.org/10.1016/B978-0-12-803581-8.03950-3
36. Liu HW, Zhang L, Wang JJ, Du XK (2008) Feasibility analysis of self-reactive spray forming TiC-TiB$_2$-based composite ceramic preforms. Key Eng Mater 368–372:1126–1129
37. Liu HW, Wang JJ, Sun XF et al (2013) Influence of cooling rate on microstructure of self-reactive spray formed Ti(C,N)-TiB$_2$ composite ceramic preforms. Adv Mater Res 631–632:348–353
38. Department of Defense Handbook (2002) Composite materials handbook, Vol. 4 Metal matrix composite MIL-HDBK-17-4A
39. Zheng X, Huang M, Ding C (2000) Bond strength of plasma-sprayed hydroxyapatite/Ti composite coatings. Biomaterials 21(8):841–849
40. Yip CS, Khor KA, Loh NL et al (1997) Thermal spraying of Ti-6Al-4V/hydroxyapatite composites coatings: powder processing and post-spray treatment. J Mater Process Technol 65(1–3):73–79
41. Shi G, Wang Z, Liang J et al (2011) NiCoCrAl/YSZ laminate composites fabricated by EB-PVD. Mater Sci Eng A 529:113–118
42. Li Y, Zhao J, Zeng G (2004) Ni/Ni$_3$Al microlaminate composite produced by EB-PVD and the mechanical properties. Mater Lett 58(10):1629–1633
43. Guo H, Xu H, Bi X, Gong S (2002) Preparation of Al$_2$O$_3$–YSZ composite coating by EB-PVD. Mater Sci Eng A 325(1–2):389–393
44. Brust S, Röttger A, Theisen W (2016) CVD coating of oxide particles for the production of novel particle-reinforced iron-based metal matrix composites. Open J Appl Sci 6:260–269
45. Chung DDL (1994) Carbon fiber composites. Butterworth-Heinemann, Boston
46. Patel RB, Liu J, Scicolone JV et al (2013) Formation of stainless steel carbon nanotube composites using a scalable chemical vapor infiltration process. J Mater Sci 48(3):1387–1395
47. Zhang G, Hu L, Hu W et al (2013) Mechanical properties of NiAl-Mo composites produced by specially controlled directional solidification. MRS Proc 1516:255–260. https://doi.org/10.1557/opl.2012.1564
48. Hu L, Hu W, Gottstein G et al (2012) Investigation into microstructure and mechanical properties of NiAl-Mo composites produced by directional solidification. Mater Sci Eng A 539:211–222
49. Gunjishima I, Akashi T, Goto T (2002) Characterization of directionally solidified B$_4$C-TiB$_2$ composites prepared by a floating zone method. Mater Trans 43(4):712–720
50. Zhang H, Springer H, Aparício-Fernandez R et al (2016) Improving the mechanical properties of Fe-TiB$_2$ high modulus steels through controlled solidification processes. Acta Mater 118:187–195
51. Zhang H, Zhu H, Huang J et al (2018) In-situ TiB$_2$-NiAl composites synthesized by arc melting: chemical reaction, microstructure and mechanical strength. Mater Sci Eng A 719:140–146
52. Yin L, Xiaonan F, Mingxu Z et al (2005) Chemical reaction of in-situ processing of NiAl/Al$_2$O$_3$ composite by using thermite reaction. J Wuhan Univ Technol Mater Sci 20(4):90–92
53. Sui B, Zeng JM, Chen P et al (2014) Fabrication of Al$_2$O$_3$ particle reinforced aluminum matrix composite by in situ chemical reaction. Adv Mater Res 915–916:788–791
54. Peng HX, Fan Z, Wang DZ et al (2000) In situ Al$_3$Ti–Al$_2$O$_3$ intermetallic matrix composite: synthesis, microstructure, and compressive behavior. J Mater Res 15(9):1943–1949
55. Singla A, Garg R, Saxena M (2015) Microstructure and wear behavior of Al-Al$_2$O$_3$ in situ composites fabricated by the reaction of V$_2$O$_5$ particles in pure aluminum. Green Process Synth 4(6):487–497
56. Lepakova OK, Raskolenko LG, Maksimov YM (2004) Self-propagating high-temperature synthesis of composite material TiB$_2$-Fe. J Mater Sci 39(11):3723–3732

57. Jin S, Shen P, Zhou D et al (2016) Self-propagating high-temperature synthesis of nano-TiC particles with different shapes by using carbon nano-tube as C source. Nanoscale Res Lett 6 (515):1–7
58. Chaubey AK, Prashanth KG, Ray N et al (2015) Study on in-situ synthesis of Al-TiC composite by self-propagating high temperature synthesis process. Mater Sci Indian J 12(12):454–461
59. Kobashi M, Ichioka D, Kanetake N (2010) Combustion synthesis of porous TiC/Ti composite by a self-propagating mode. Materials 3:3939–3947
60. Pramono A, Kommel L, Kollo L et al (2016) The aluminum based composite produced by self-propagating high temperature synthesis. Mater Sci 22(1):41–43

Chapter 4
Fabrication and Characterization of Composites

4.1 Synthesis and Characterization of Mg/TiC Composites

In the last years, magnesium and its alloys have been used in the fabrication of MMC. Most of them use SiC- [1–6] and TiC-like [7–12] reinforcement. These composites are attractive candidates for automotive and aerospace applications due to their low density, wear resistance, and lower coefficients of thermal expansion, and they are light materials and have good mechanical properties.

A number of manufacturing techniques have been developed to produce metal matrix composites (MMCs). When high levels of reinforcement are required, the infiltration of liquid metal into porous ceramic preforms is preferred since it can yield near-net-shape components with high stiffness and enhanced wear resistance [13]. The driving force for infiltration of the liquid metal is dictated by the affinity between the metal and ceramic couples in terms of wettability. When good wettability exists in a given metal-ceramic system, and under adequate conditions of temperature and atmosphere, the liquid metal may be drawn into the ceramic preform simply by capillarity.

Sessile drop experiments revealed a non-wetting/wetting transition in the Mg/TiC system between 800 and 850 °C in Ar, without formation of new compounds at the interface, suggesting TiC is a stable reinforcement for Mg composites [14]. Oxide layers in metals, such as Al and Mg, represent a mechanical obstacle that has to be overcome before the inherent wettability of the system is exhibited. While Al oxidizes forming a compact and passivating layer, Mg forms a porous and non-protective oxide [8, 14]. Moderate mechanical and physical properties of Mg require the incorporation of suitable ceramic particles can compensate for many of these limitations. In the last years, many studies using Mg and Mg alloys have been used to fabricate MMC. TiC is a hard and stiff transition metal carbide with low density and has recently been dispersed in an Mg matrix via a master Al/TiC composite [3].

© Springer Nature Switzerland AG 2018
A. Contreras Cuevas et al., *Metal Matrix Composites*,
https://doi.org/10.1007/978-3-319-91854-9_4

Dong et al. [9] fabricated and characterized the TiC_p reinforced magnesium matrix composites by in situ reactive infiltration process. They found that the smaller elemental particle size and a processing temperature above 700 °C were beneficial to synthesizing Mg/TiC_p composites. Zhang Xiuqing et al. [15] characterized the damping behavior of 8 wt.% TiC reinforced magnesium matrix composites. Damping capacity of magnesium matrix composites is generally higher due to the addition of TiC particulates than that of AZ91 magnesium alloy. Jiang et al. [16] conducted a process via self-propagating high-temperature synthesis reaction in a TiC particulate reinforced magnesium matrix composite. The results revealed that composites have higher properties (such as hardness, UTS, and wear resistance) compared to those of the unreinforced magnesium alloy. Balakrishnan et al. [17] performed an experiment to synthesize Mg-AZ31/TiC composites. Four different volume fractions of TiC particles (0, 6, 12, and 18 vol.%) were used. The SEM analysis indicated that there was no interfacial reaction between the magnesium matrix and the TiC particles.

Gu et al. [18] investigated the microstructure and mechanical properties of transient liquid phase (TLP) bonded TiC reinforced magnesium metal matrix composite ($TiC_p/AZ91D$) joints using aluminum interlayer at 460 °C. The microstructure revealed the presence of α-Mg and $Al_{12}Mg_{17}$ compounds. The increase of $Al_{12}Mg_{17}$ compound and aggregation of TiC particulates were the main reason for affecting the mechanical properties of joints. The joint shear strength above 58 MPa was obtained at the bonding temperature of 460 °C. Anasori et al. [19] studied the mechanical properties of Mg matrix composites fabricated by pressureless melt infiltration of Mg and Mg alloys (AZ61 and AZ91) into porous preforms of TiC and Ti_2AlC with 50% vol. The best properties were obtained when AZ61 was reinforced with Ti_2AlC particles.

Kaneda and Choh [20] fabricated a particulate reinforced magnesium matrix composites using a spontaneous infiltration. Pure magnesium was used as matrix and SiO2 and SiC particles (SiC_p, 1.2, 2, 3, 4 and 8 μm) as reinforcement. MgO and Mg_2Si reaction products were observed. More recently, several reviews about fabrication and characterization of magnesium matrix composites have been carried out [1, 21].

Our research group studied the feasibility of pressureless infiltration of molten pure Mg into TiC preforms [8]. The effect of temperature on both, infiltration kinetics and mechanical properties, was investigated.

4.1.1 Experimental Conditions

Porous preforms were prepared by uniaxially pressing 18.5 g of TiC powder with an average particle size of 1.2 μm at 8 MPa in a rectangular die to form green bars ~65 × 10 × 10 mm in size. The preforms were sintered at 1250 °C for 1 h in a tube furnace under flowing Ar, yielding porous preforms with 56% of theoretical density.

A thermogravimetric technique [22] was used to follow the infiltration behavior at 850, 900, and 950 °C under an Ar atmosphere. The weight gained by the preforms, as a result of infiltration of liquid Mg into the TiC preforms, was continuously monitored to obtain characteristic infiltration profiles. The resulting composites were observed in a scanning electron microscope (SEM) attached to an energy-dispersive X-ray spectroscopy (EDX) system. X-ray diffraction (XRD) was used to identify the phases present.

Some mechanical properties of the Mg/TiC composites like elastic modulus, hardness, and tensile strength were evaluated. Fracture surfaces were observed using SEM.

4.1.2   Microstructural Characterization of the Composites

Green preforms were sintered at 1250 °C for 1 h to produce TiC preforms with ~44% interconnected porosity. Figure 4.1a shows a SEM image of the TiC preforms. Pressureless melt infiltration of molten Mg into these preforms was successfully achieved at temperatures of 850, 900, and 950 °C under flowing Ar. Figure 4.1b shows a typical micrograph of the resulting composites. The dark phase corresponds

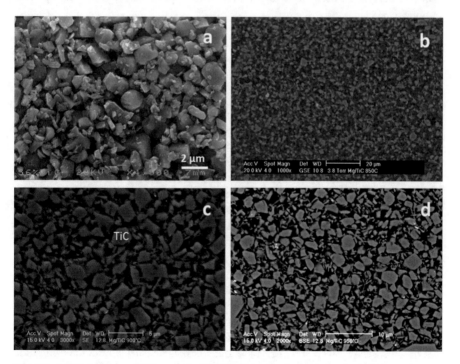

Fig. 4.1 (**a**) Morphology of TiC preform, Mg/TiC composite infiltrated at (**b**) 850 °C, (**c**) 900 °C, and (**d**) 950 °C

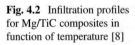

Fig. 4.2 Infiltration profiles for Mg/TiC composites in function of temperature [8]

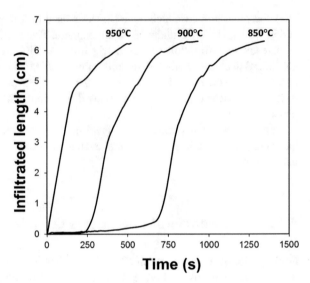

to Mg and the gray phase to TiC particles. Although the full length of the bars was infiltrated, some porosity was observed in the microstructure. The infiltrated composites that reach densities of about 97% with approximately 3% porosity were produced.

4.1.3   Infiltration Kinetics

Figure 4.2 shows the infiltration profiles obtained for Mg/TiC composites infiltrated at different temperatures. Temperature dependence of the infiltration rate is clear; the higher the temperature, the faster the rate of infiltration. There is an incubation period prior to stabilizing into parabolic-type infiltration curves. This incubation period also seems to be temperature dependent, indicating that infiltration of liquid Mg in TiC preforms is preceded by a thermally activated process. The presence of an incubation period has commonly been observed during the infiltration of TiC by Al and Al alloys and has been ascribed to the unstable wetting behavior in these systems [22, 23]. The incubation period in the infiltration of Mg in TiC can be also related to the time required for the Mg melt, in contact with the TiC preform, to achieve the threshold contact angle for infiltration to proceed at a given temperature.

Kinetic studies were carried out on the infiltration profiles (dl/dt) and the incubation period ($1/t_0$) because these are temperature dependent. The corresponding rate constants (k) can be determined at each stage at a given temperature. Results of the Arrhenius analysis are shown in Tables 4.1 and 4.2. The results revealed high activation energies, suggesting that mass transfer mechanisms control the infiltration process instead of viscous flow for which activation energies fall in the order of about 10 kJ/mol [23].

Table 4.1 Activation energy estimated from the incubation periods

$T_{Inf.}$ (°C)	$1/T$ (K)	Incubation time (t_0)	$1/t_0$ (1/s)	Ln ($1/t_0$)	Slope ($-E_a/R$)	E_a (kJ/mol)
850	0.000890	680	0.0014	−6.52	−60,387	503
900	0.000852	240	0.0041	−5.48		
950	0.000817	8	0.125	−2.07		

Table 4.2 Activation energy estimated from the steady part of the infiltration profiles

$T_{Inf.}$ (°C)	$1/T$ (K)	dl/dt (cm/s)	Ln (dl/dt)	Slope ($-E_a/R$)	E_a (kJ/mol)
850	0.00089	0.000700	−7.264	−47,054	392
900	0.00085	0.006402	−5.051		
950	0.00081	0.028609	−3.554		

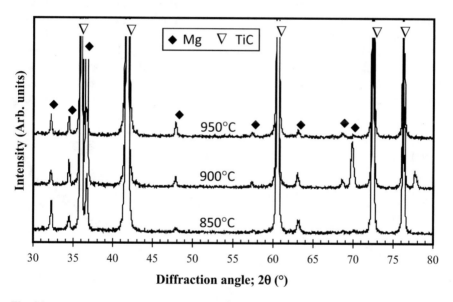

Fig. 4.3 XRD patterns for the infiltrated Mg/TiC composites [8]

4.1.4 Interfacial Reactions Between Matrix and Reinforcement

SEM examination and EDX analysis performed on the different samples did not show evidence of an additional phase, other than TiC and Mg. This was also corroborated by XRD, as shown in Fig. 4.3. This might be expected as the formation of Mg carbide is thermodynamically unfavorable. Also, Mg and Ti do not form intermetallics, and the solubility of Ti in Mg is very limited under equilibrium conditions [24]. Thus, TiC can be considered stable in pure Mg melts.

Non-oxide ceramics exhibit a superficial oxide layer; the level of surface oxida-tion for TiC varies from an oxycarbide TiC_xO_{1-x} to Ti suboxides and to polymorphic TiO_2 [25]. The high affinity of Mg for oxygen suggests that during infiltration, an oxycarbide layer on the surface of the TiC preform can be readily reduced by the advancing liquid magnesium front and that this surface reaction is the possible driving mechanism for the infiltration process.

4.1.5 Mechanical Properties

Tensile specimens from the composites were machined according to ASTM E8. Special care had to be taken machining the specimen because they are brittle. Figure 4.4 shows the stress-strain profiles obtained from Mg/TiC composites. In this figure, it can be observed that profiles showed only an elastic region (straight line), and little deformation is exhibited. Most of the tests performed at different infiltration temperatures showed strain values around 8–10%. Table 4.3 summarizes the mechanical properties for the composites. As expected, a significant increase in the elastic modulus is observed with regard to pure Mg ($E = \sim 45$ GPa). An increase in the mechanical properties is also observed by increasing the infiltrating

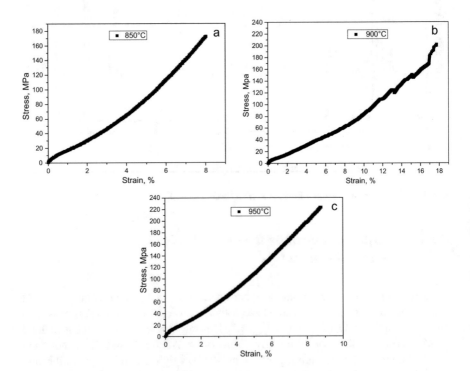

Fig. 4.4 Stress-strain profiles obtained from Mg/TiC composites

Table 4.3 Mechanical properties of the Mg/TiC composites

Temperature (°C)	E (GPa)	$E_{Halpin-Tsai}$ (GPa)	$E_{Boccaccini\ et\ al}$[a] (GPa)	UTS (MPa)	Vickers hardness	Strain (%)
850	123	138 ($s = 1$)	123–136	172	183–191	8–9
900	130	153 ($s = 1.5$)	123–137	200	194–197	16–18
950	136		122–135	233	205–212	8–10

[a]Aspect ratio of the pores, 1; orientation factor ($\cos^2\alpha_d$), 0.31

temperature, which could be attributed to better infiltration and consequently a strong interface between Mg and TiC particles. Elasticity modulus measurements were obtained and were also compared with a predicted value using the Halpin-Tsai model [26, 27] for 1 and 1.5 aspect ratios (s) of the reinforcing particles. The values calculated from the model, which do not account for defects such as porosity, provide a slightly overestimated but reasonable theoretical prediction. Taking account that porosity levels in the composites were low and comprised mostly of closed porosity, Boccaccini and Fan [28] use a model that can be used to predict the elasticity modulus and estimate a theoretical value by taking into account this factor.

The ultimate tensile strength (UTS) and hardness values were also seen to be related to the infiltrating temperature. While tensile testing evaluates the bulk of the specimens, hardness measurements, performed in confined, pore-free, volumes of the samples, provide an insight into the integrity of the particle-matrix interface which is directly related to the degree of intimate bonding achieved during processing. This aspect, therefore, has to be considered in relation to the mechanical response of the composites, since a good particle-matrix interfacial bonding improves load transfer, increasing stiffness and delaying the onset of particle-matrix decohesion.

No necking was observed during tensile testing specimens, suggesting that the composites failed rapidly after yielding. This observation, however, does not necessarily mean that the composites fully lack ductility. Figure 4.5 shows SEM fractographs of composites processed at different temperatures. Particle-matrix de-bonding can be appreciated (Fig. 4.5b–d), accompanied by scarce formation of micro-dimples. These features indicate that the failure mechanism is related to matrix voiding, decohesion of TiC particles, and the subsequent growth and coalescence of the corresponding micro-voids, rather than on shearing of the Mg phase impeded by the TiC particles. Some fractured TiC particles are also observed, which is indicative of good particle-matrix bonding.

Pressureless-infiltrated TiC preforms with molten Mg were obtained at 850 °C under an inert Ar atmosphere. Prior to steady-state infiltration, characteristic infiltration profiles exhibited an incubation period, which was temperature dependent and related to the dynamic wetting behavior previously observed in the Mg/TiC system. High activation energy values suggest that the infiltration process is controlled by an interfacial mass transfer mechanism. Composites are free from reaction products. The mechanical properties of the composites exhibited a trend to increase with infiltrating temperature, which can be attributed to a better and complete infiltration of the preforms. Decohesion of TiC particles in the fracture specimens was observed.

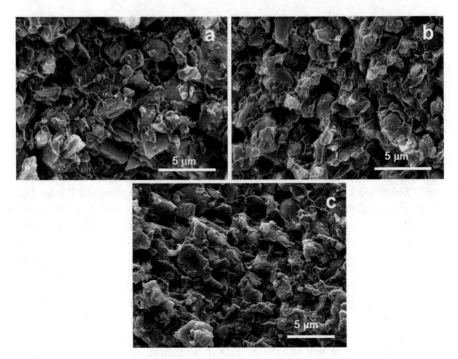

Fig. 4.5 Fractographs of the composites infiltrated, (**a**) 850 °C, (**b**) 900 °C, and (**c**) 950 °C [8]

4.2 Synthesis and Characterization of Mg-AZ91/AlN Composites

Currently, the most widely used materials as reinforcement are Al_2O_3, SiC, TiC, and graphite, which have been used in aluminum, magnesium, and its alloys with the purpose of improving its mechanical properties such as the module of elasticity, hardness, and wear resistance, among others.

In the elaboration of metal matrix, composites with high contents of reinforcement generally used three routes: powder metallurgy, [29, 30] in which ceramic particles and metal matrix are mixed, pressed, degassed, and sintered; fabrication in situ at high temperatures [31–34], in which a reaction of a mixture of powders and/or gas take place to produce a composite material; and, maybe one of the most used methods, the infiltration technique [35–37], in which a liquid metal penetrates in a porous preform to fill in all of the interstices of the same kind, for which in some cases, it is required to apply some external force that could be mechanic, like the case of squeeze casting [38–40] or applying gas pressure [41]. One of the advantages of using infiltration technique is that the infiltrate preforms have dimensions and forms close to the piece required. In case of infiltration without external pressure, the driving force for infiltration of the liquid metal is dictated by the affinity between the metal and ceramic couples in terms of wettability. Therefore, it is important to have a good wettability to obtain strong links between the matrix and reinforcement [42].

The nature of the technique utilized in the fabrication of composite materials has a marked effect over the properties of the material. The infiltration by capillarity without external pressure in ceramic preforms is an attractive technique to manufacture materials with high ceramic content [35, 36]. However, the temperatures relatively high that are required to fabricate this compound by this technique could generate products of reaction in the interface. A composite with magnesium alloy AZ91E matrix and aluminum nitride (AlN) offers good characteristics for its use as a functional material, particularly in electronic packaging industry.

It has been reported in the literature the elaboration of MMC, using AlN as reinforcement and different processing routes [30, 31, 34, 35, 38, 39, 41, 43–45]. However, a few of these studies have used magnesium and its alloys; nevertheless the use of these elements has increased in the last few years in the automotive industry, due to its light weight. As a result, the alloys of magnesium offer a specific resistance compared to the conventional alloys; also the AZ91E Mg alloy have an excellent castability, good machinability, and good corrosion resistance [46].

On the other hand, AlN polycrystalline has a thermal conductivity of 80–200 Kw/m K and a coefficient of thermal expansion of $4.4 \times 10^{-6}/°C$ [47]. These two properties make AlN an excellent material for electric circuits of high density compared with other ceramic substrates that have low coefficients of thermal conductivity and high thermal expansion. As a result, the combination of the mechanical properties of AlN with the magnesium and its alloys gives an origin to an attractive compound for electronic and structural applications.

This research presents the synthesis and characterization of metal matrix composites (MMC) of magnesium alloy AZ91E reinforced with AlN (49 vol.%). Molten AZ91E magnesium alloy was pressureless infiltrated at 900 °C into AlN preforms sintered at 1450 °C with porosities around 51 vol.%.

4.2.1   Experimental Conditions

Table 4.4 shows the chemical composition of AZ91E magnesium alloy (Thomson Aluminum Casting Co. USA) used in the fabrication of the composite material. The reinforced material was powder of AlN (Aldrich Chemical Co. USA) with an average in particulate size of 1.38 μm.

AlN preforms were fabricated with 12 g of powders in a rectangular metal die with dimensions of $6.5 \times 1 \times 1$ cm. The Al powders were compacted using uniaxial pressure (15 MPa). Preforms were sintered at temperature of 1450 °C during 1 h in a nitrogen atmosphere (99.99%). Density and porosity of the sintered preforms as well as the composite were evaluated using the Archimedes method described in ASTM C20-00 [48]. Finally, the synthesized preforms were placed in contact with small

Table 4.4 Chemical composition of AZ91E alloy (wt.%)

Mg	Al	Zn	Mn	Si	Fe	Cu	Ni
90	8.1–9.3	0.4–1	0.17–0.35	0.2max	0.005max	0.015max	0.001max

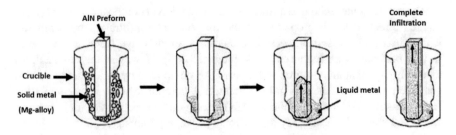

Fig. 4.6 Schematic representation from infiltration process of the Mg-AZ91E/AlN composite [47]

pieces of AZ91E alloys in a graphite crucible inside a horizontal tubular furnace at temperature of 900 °C for 10 min in argon atmosphere. Figure 4.6 shows a schematic representation of the infiltration process.

The obtained composite materials were microstructurally characterized using X-ray diffraction and scanning and transmission electronic microscopy. In addition, some mechanical (hardness and elastic modulus) and thermal properties (coefficient of thermal expansion, CTE) were evaluated. The structural and morphological characteristics and the products of reaction in the interface were analyzed using transmission electron microscopy (TEM).

4.2.2   Microstructural Characterization of the Composites

After the bars were infiltrated, several samples were prepared to be observed through SEM. Figure 4.7 shows a SEM micrograph typical of the composite material Mg-AZ91E/AlN. A heterogeneous size of AlN particles can be observed, which is approximately 1–5 μm. The composites produced have 51% of metal matrix and 49% of reinforcement (AlN). SEM observations at high magnifications revealed that apparently interfacial reaction does not exist between the reinforcement and the matrix. The bulk density of the composites measured according to Archimedes method was around 98% and around 2% of close porosity approximately.

Figure 4.8 shows an elemental distribution mapping of the main elements of the composite. It can be observed that magnesium matrix was homogeneously distributed along the composite.

Figure 4.9 shows a diffractogram from the metal matrix composites produced a 900 °C. Through XRD it can be identified the AlN, Mg, and $Mg_{17}Al_{12}$ phases. Thermodynamics says that formation of MgO and the spinel $MgAl_2O_4$ is possible; however these phases were not detected by XRD, which generally forms as interfacial reaction products.

The Mg-AZ91/AlN composite system has Al-Mg-Zn-N-Si-O like the main alloying elements; thus the possible reaction products that can form at the interface of AlN and Mg-AZ91 alloy could be $Mg_{17}Al_{12}$, Mg_2Si, Mg_2Al_3, Al_2O_3, $MgAl_2O_4$, and MgO mainly. Because the oxide bonds (Al_2O_3, $MgAl_2O_4$, and MgO) are significantly stronger than $Mg_{17}Al_{12}$, Mg_2Si, and Mg_2Al_3 compounds, therefore,

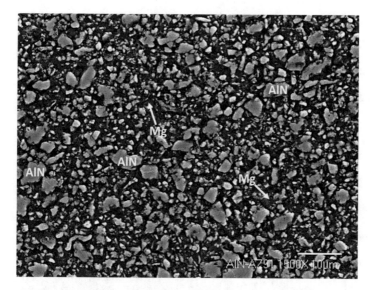

Fig. 4.7 SEM micrograph of the Mg-AZ91/AlN composite

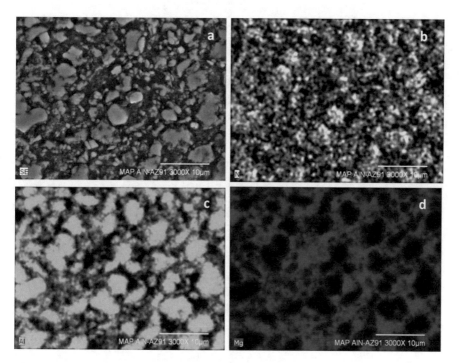

Fig. 4.8 (**a**) SEM micrograph of the Mg-AZ91E/AlN composite and elemental chemical mapping of (**b**) nitrogen, (**c**) aluminum, and (**d**) magnesium

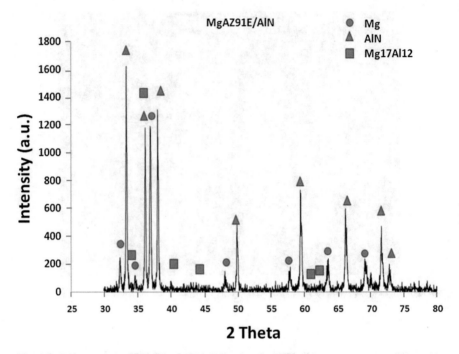

Fig. 4.9 XRD spectrum of the Mg-AZ91/AlN composites [47]

oxide phases should form in preference to other possible compounds, until all of the available oxygen is consumed. Among the possible oxides, the formation of free energy for MgO is lower than that of $MgAl_2O_4$ and Al_2O_3. The study on thermodynamic stability of Al-Mg oxides in Al-Mg alloys shows that the formations of Al_2O_3, $MgAl_2O_4$, and MgO are competitive processes, with preferential MgO formation for high Mg content [49–53].

According to free Gibbs energy as shown in the following reaction, the formation of Mg and $MgAl_2O_4$ is possible:

$$Mg_{(l)} + 1/2 O_{2(g)} = MgO_{(s)}$$
$$\Delta G_{(900°C)} = -473 \, kJ \tag{4.1}$$

$$3Mg_{(l)} + 4Al_2O_{3(s)} = 3MgAl_2O_{4(s)} + 2Al$$
$$\Delta G_{(900°C)} = -256 \, kJ \tag{4.2}$$

The formation of $MgAl_2O_4$ can be also feasible in solid state according to the following reaction:

$$MgO_{(s)} + Al_2O_{3(s)} = MgAl_2O_{4(s)}$$
$$\Delta G_{(900°C)} = -44 \, kJ \tag{4.3}$$

However, MgO is more stable thermodynamically than Al_2O_3. Therefore, Mg can reduce the Al_2O_3 according to the following reaction:

$$3Mg_{(l)} + Al_2O_{3(s)} = 3MgO_{(s)} + 2Al$$
$$\Delta G_{(900°C)} = -123\,kJ \tag{4.4}$$

As can be observed, considering the Gibbs free energy, the reaction that gives origin to formation spinel is feasible to occur. However, the degree of reaction depends on many conditions such as temperature, time, microstructure of reinforcement, and chemical composition, among others.

Al_2O_3 is thermodynamically stable in contact with Al-pure; however, in contact with Mg-liquid, it tends to form MgO according to the Eq. (4.4) or the spinel $MgAl_2O_{4(s)}$ according to Eqs. (4.2) and (4.3), being more feasible than Eq. (4.2). In a system that has enough oxygen, this spinel can be formed reacting with Mg and Al according to the following reaction:

$$Mg_{(l)} + 2Al_{(l)} + 2O_{2(g)} = MgAl_2O_{4(s)}$$
$$\Delta G_{(900°C)} = -1808\,kJ \tag{4.5}$$

The spinel formation can be feasible even in solid state as shown thermodynamically in Eq. (4.3). In the Mg-AZ91/AlN system, this spinel was not detected.

4.2.3   TEM Observations

Thermodynamic considerations show that the most stable oxide will be MgO. In order to clarify the possible interfacial reaction products and phases present on the Mg-AZ91/AlN composite, some samples were prepared and observed through transmission electron microscopy (TEM).

Figure 4.10a shows a high resolution transmission electron microscopy (HRTEM) image of the MgO precipitate which is embedded in the magnesium matrix. Figure 4.10b shows a HRTEM image of an MgO precipitate. The planar distances, which were measured directly from the image, correspond to 2.473, 1.972, and 1.488 Å (JCPDS card number 65-0476). All the diffraction patterns obtained from the precipitates correspond closely to the parameters defined by the cubic MgO phase ($a = 0.42$ nm). Figure 4.10c shows the fast Fourier transform (FFT) obtained from HRTEM image of the MgO precipitate; this precipitate is oriented along the $[1\,-1\,0]$ zone axis. After careful indexing, this phase was identified as MgO. From the FFT the planar distances for this phase belong to (111), (002), and (220) planes, observed in the $[1\,-1\,0]$ zone axis.

Through bright-field scanning transmission electron microscopy (BF-STEM) image, some MgO precipitates were observed as shown in Fig. 4.11. Figure 4.11a shows a BF-STEM image where it can see a precipitate of MgO in the grain

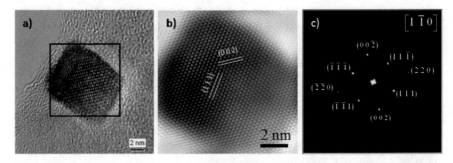

Fig. 4.10 (**a**) HRTEM image which corresponds to the MgO precipitate embedded in the magnesium matrix, (**b**) HRTEM image of the MgO showing the interplanar distances corresponding to (111) and (002) planes, and (**c**) FFT obtained from HRTEM image oriented in the [1 −1 0] zone axis [47]

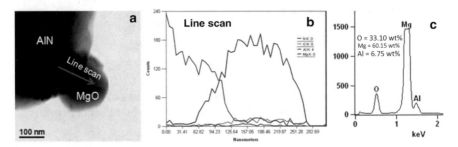

Fig. 4.11 (**a**) BF-STEM image showing the formation of some MgO precipitates, (**b**) line scan through AlN and MgO interface, and (**c**) EDS spectrum obtained from the MgO precipitate [47]

boundaries of AlN particle. A line scan through AlN and MgO interface was carried out to determine the elemental composition, and the results are shown in Fig. 4.11b. Additionally, at this precipitate an EDS spectrum was performed whose results are shown in Fig. 4.11c.

Figure 4.12a shows a high-angle annular dark-field (HAADF) STEM image showing the presence of some precipitates of several sizes, which precipitates preferentially in the AlN grain boundaries. Figure 4.12b shows the selected area electron diffraction (SAED) pattern obtained from the region marked with a red circle of Fig. 4.12a. To determine the phase of these precipitates, the SAED pattern was indexed as is shown in Fig. 4.12b. The crystallographic analysis of the planar spacing correspond to (5 1 0), (1 −1 6), and (6 0 6) planes of the $Mg_{17}Al_{12}$ phase observed in [1 −5 −1] zone axis (JCPDS card number 01-1128). Additionally, an EDS microanalysis at these precipitates shown in Fig. 4.12c was carried out. The results of the stoichiometric analysis are consistent with the $Mg_{17}Al_{12}$.

The same phase was found by Zheng et al. [54], where they reported a red parameter of 1.056 nm. They studied the behavior of a $SiC_w/AZ91$ compound and found precipitation $Mg_{17}Al_{12}$ phase at the interface.

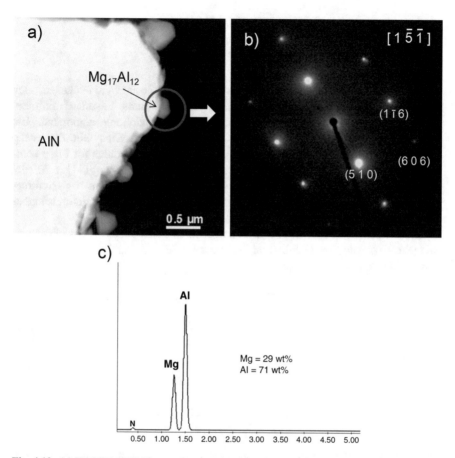

Fig. 4.12 (**a**) HAADF-STEM image showing the $Mg_{17}Al_{12}$ precipitates, (**b**) SAED pattern of the precipitate shown in Fig. 4.12a, and (**c**) EDS spectrum of the particle shown in image 4.12a [47]

A strongly bonded interface is a requirement to have a good mechanical and structural behavior in a compound. The nature and quality of the interface (morphology, chemical composition, reactivity, strength, and adhesion) are determined by intrinsic factors of both the reinforcement material and the matrix (chemical composition, crystallography, and defect content) as well as by intrinsic factors related to process (time, temperature, pressure, atmosphere, etc.).

Moderate interfacial reactions between matrix and reinforcement sometimes are good and improve the resistance of the interface and load transfer. However, excessive chemical reaction degraded the reinforcement and resistance of the compound [51, 52]. To control the extent of interfacial reaction, it is practical to reduce the processing time of the compounds, apply metal coatings and add alloying elements in the matrix.

4.2.4 *Mechanical Characterization*

4.2.4.1 Elastic Modulus

Some mechanical properties of the Mg-AZ91/AlN composite were evaluated, such as elastic modulus and hardness, which directly depends on the content of reinforcement. The elastic modulus was measured through GrindoSonic equipment. The average value of elastic modulus obtained was 133 GPa. Meanwhile the elastic modulus of the Mg-AZ91 alloy was 45 GPa. The elastic modulus for these composites is similar to the elastic modulus reported by Lai and Chung [44] in Al/AlN system with 58% of reinforcement (144 GPa). However the volumetric reinforcement that they used is higher; in addition the elastic modulus of aluminum is higher than AZ91 alloy.

Theoretical evaluations of elastic modulus were carried out by means of mixture rule (Eq. 4.6), using the Halpin-Tsai [27] equation (Eqs. 4.7 and 4.8) and using the Hashin-Shtrikman equation (Eq. 4.9) [26, 43]:

$$E_c = {}^c V_m E_m + V_r E_r \qquad (4.6)$$

$$E_c = \frac{E_m(1 + 2SqV_r)}{1 - qV_r} \qquad (4.7)$$

$$q = \frac{(E_r/E_m) - 1}{(E_r/E_m) + 2S} \qquad (4.8)$$

$$E_c = E_m \cdot \frac{E_m \cdot V_m + E_r(V_r + 1)}{E_r \cdot V_m + E_m(V_r + 1)} \qquad (4.9)$$

where E_c is the elastic modulus of composite (GPa), V_m is the percentage (%) of volumetric matrix, V_r is the percentage (%) of volumetric reinforcement, E_m is the matrix modulus (GPa), E_r is the reinforcement modulus (GPa), and S is the aspect ratio.

The rule of mixture is the simplest model. Halpin-Tsai model is a mathematical model for the prediction of elasticity of composite material based on the geometry and orientation of the reinforcement and the elastic properties of the reinforcement and matrix. The model is based on the self-consistent field method although often considers to be empirical.

The results obtained from elastic modulus measured according to mixture rules and Hashin-Shtrikman and Halpin-Tsai equations for different aspect ratios are shown in Fig. 4.13. Several calculations were included in order to observe the effect of volumetric content of the reinforcement. It is clear that increasing the reinforcement content, the elastic modulus increases as is shown in Fig. 4.13.

The elastic modulus obtained with the Halpin-Tsai equation is closer to the elastic modulus measured with the GrindoSonic equipment. The difference in the assessment with the different methods is attributed to Halpin-Tsai model, and a mixture rule does not consider the porosity of the composites. Good approach with an aspect ratio of 2 was obtained.

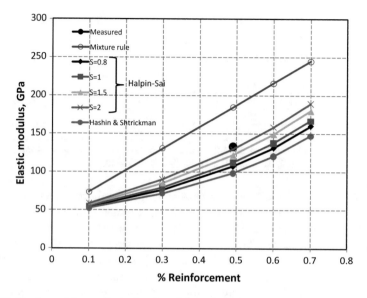

Fig. 4.13 Elastic modulus results in function of the volumetric reinforcement content [47]

4.2.4.2 Hardness Measurements

The hardness of the composites was measured in several points of the bar obtaining an average value of 24 HRC (260 HV). It is important to mention that important variations in the values of hardness measured along the bar were not observed, which is an indication that the composites are homogeneous and thus have homogeneous distribution of particle reinforcement.

4.2.5 Thermoelectric Characterization

A thermoelectric characterization of Mg-AZ91/AlN composite through coefficient of thermal expansion (CTE) and electric resistivity measurements was carried out.

4.2.5.1 Coefficient of Thermal Expansion (CTE)

Aluminum nitride has a hexagonal crystal structure and is a covalent-bonded material. AlN is stable to very high temperatures in inert atmospheres. In air, surface oxidation occurs above 700 °C. The coefficient of thermal expansion for AlN is 4.5×10^{-6} °C^{-1}. Meanwhile, for Mg alloy it was considered as 2.48×10^{-5} °C^{-1}.

The thermal expansion coefficient measured at the Mg-AZ91/AlN composite was in the order of 9.53×10^{-6} °C^{-1}. Zhang et al. [39] reported values

Table 4.5 CTE assessments in function of the volumetric reinforcement content

Reinforcement (%)	Matrix (%)	CTE mixture rule ($°C^{-1}$)	CTE Hashin-Shtrikman ($°C^{-1}$)	CTE measured ($°C^{-1}$)
0.1	0.9	$6.5e^{-6}$	$5.13e^{-6}$	
0.3	0.7	$10.6e^{-6}$	$6.82e^{-6}$	
0.49	**0.51**	**$14.4e^{-6}$**	**$9.09e^{-6}$**	**$9.53e^{-6}$**
0.6	0.4	$16.7e^{-6}$	$1.08e^{-5}$	
0.7	0.3	$18.7e^{-6}$	$1.29e^{-5}$	

It is the percentage of reinforcement (0.49%) and matrix (51%) in the composites fabricated. The other percentages (0.1, 0.3, 0.6 and 0.7) are only theoretical calculations.

of $11.2 \times 10^{-6} °C^{-1}$ for Al/AlN composite with 50% reinforcement, and Lai [44] obtained values of $10.16 \times 10^{-6} °C^{-1}$ for Al/AlN composite with 54.6% reinforcement. Additionally, some calculations of CTE were carried out with the mixture rule, and using the Hashin-Shtrikman [26, 27, 43] equation, the results are shown in Table 4.5. A very good agreement between the CTE measured in the composite and the CTE calculated using the equation of Hashin-Shtrikman was obtained.

Most current applications are in the electronics area where heat removal is important. Aluminum nitride and silicon carbide are good for transferring heat. Aluminum nitride is used in packages for semiconductors that emit high volumes of heat but must avoid accumulating heat internally. The composites fabricated with AlN as reinforcement can be used as structural and functional materials; thus these composites can be used as electronic packing material where they are the required materials with low thermal expansion coefficient and high thermal conductivity.

4.2.5.2   Electrical Resistivity (ER)

The electric resistivity measured in the Mg-AZ91/AlN composite was in the order of 45.9×10^6 Ω cm at 25 °C. Thermal conductivity of materials is temperature dependent. In general, materials become more conductive to heat as the average temperature increases. Correspondingly materials of high thermal conductivity are widely used in heat sink applications (like computer fan), and materials of low thermal conductivity are used as thermal insulation. Considering an electrical resistivity of 4.42×10^{-6} and 1×10^{12} Ω cm for Mg alloy and AlN, respectively, it was evaluated the ER using the mixture rules, and the results are shown in Table 4.6.

Fabrication and characterization of AZ91E/AlN composite were carried out. AlN, Mg, and $Mg_{17}Al_{12}$ phases were detected through X-ray diffraction. Additionally, TEM results confirmed the presence of small precipitates of MgO and $Mg_{17}Al_{12}$. Results of mechanical characterization of the composites indicate average values of the elastic modulus of 133 GPa and a hardness of 24 HRC (260 HV). It is clear that increasing the reinforcement content, the elastic modulus increases. The coefficient of thermal expansion (CTE) was $9.53 \times 10^{-6} °C^{-1}$, and the electrical resistivity of the composite was 45.9×10^6 Ω cm. A very good agreement between the CTE measured in the composite and the CTE calculated using the equation of Hashin-

Table 4.6 Electrical
resistivity (ER) of the
composite in function of the
volumetric reinforcement
content

Reinforcement (%)	Matrix (%)	ER-mixture rule (Ω cm)	ER-measured (Ω cm)
0.1	0.9	$10.0e^{10}$	
0.3	0.7	$30.0e^{10}$	
0.49	**0.51**	**$49.0e^{10}$**	**$45.9e^{6}$**
0.6	0.4	$60.0e^{10}$	
0.7	0.3	$70.0e^{10}$	

It is the percentage of reinforcement (0.49%) and matrix (51%) in the composites fabricated. The other percentages (0.1, 0.3, 0.6 and 0.7) are only theoretical calculations.

Shtrikman [26] was obtained. The composite fabricated shows very attractive mechanical, thermal, and electrical properties. The processing route permits to have control of vol.% reinforcement allowing to obtain composites with a very high volume fraction of reinforcement, making the composite a candidate to be used.

4.3   Synthesis and Characterization of Mg/AlN Composites

A composite with continuous interconnected ceramic and metal phases has been fabricated from sintered porous particulate AlN preforms infiltrated with pure magnesium [35]. The preforms were fabricated by pressureless infiltration in argon in the temperature range of 870–960 °C obtaining composites with 48 vol.% AlN.

The use of AlN-like reinforcement in composites by pressure melt infiltration [43], vacuum liquid metal infiltration [44], melt stirring [55], and squeeze casting [56] has been reported, which include aluminum and magnesium alloys as matrices.

The common method for the fabrication of MMCs with high content of reinforcement phase is the infiltration of molten metal into a preform containing continuous open porosity. This pressureless infiltration method offers the advantage of producing complex shapes with low residual porosity at relatively low cost and, for the fabrication of magnesium MMCs, a low process temperature [8, 9, 20, 22, 23, 36]. The thermal conductivity of sintered polycrystalline AlN is typically 80–200 W/m K, with a coefficient of thermal expansion (CTE) of 4.4 $\times$ 10^{-6}/°C, [47]. These two properties make AlN an excellent material for high-power, high-density circuits compared with other ceramic substrates that generally exhibit lower thermal conductivities and/or higher CTE [47, 57].

4.3.1   Experimental Conditions

The reinforcing material employed in the fabrication of Mg-based MMCs was AlN powder with 1.38 μm average particle size (Aldrich Chemical Co.). In the case of the matrix, electrolytic Mg was employed to carry out the infiltration. Rectangular compacts of AlN were prepared by die pressing to form green bars about

$6.5 \times 1 \times 1$ cm in size. The preforms were sintered for 1 h at 1500 °C under nitrogen, leading to compacts with densities of 48 vol.% and residual porosity of 52 vol.% that provided green strength for easy manipulation during the infiltration process. Pressureless infiltration was conducted at the temperatures of 870, 900, 930, and 960 °C under argon by a metal-ceramic contact technique using a tubular furnace and graphite crucibles to contain the metal.

After obtaining the Mg/AlN composites, they were evaluated through some mechanical tests. Hardness measurement was conducted on a Rockwell hardness tester (Wilson) using a load of 100 kg in Rockwell B scale. Young's modulus was measured with a GrindoSonic MK5 (J. W. LEMMENS N. V.). The tensile properties of the composites were assessed by the shear punch test (SPT). Four tests were carried out for each infiltration temperature, and an average of UTS was obtained. This technique has been successfully used by some authors in the mechanical characterization of infiltrated coated ceramics [58]. Polished samples were examined by X-ray diffraction (SIEMENS D5000) and scanning electron microscopy (SEM, JEOL JSM-6400) coupled with an energy-dispersive spectroscopy (EDS) analyzer system. The interface of the Mg/AlN composites was characterized with a field-emission gun transmission electron microscope (FEG-TEM, Philips Tecnai F20) coupled with an EDS analyzer system. The CTE values of Mg/AlN composites were measured using a dilatometer (Theta Dilatronic) operated at a heating rate of 5 °C/min.

4.3.2   Microstructural Characterization of the Composites

Samples of Mg/Al composites were selected to be polished and prepared to be observed through scanning electron microscopy (SEM) and transmission electron microscopy (TEM). Figure 4.14 shows a SEM micrograph of a polished Mg/AlN composite infiltrated at 930 °C. The dark phase surrounding the AlN particles is the magnesium matrix. All the Mg/AlN composites infiltrated at different temperatures show a uniform distribution of AlN particles in an interconnected matrix. Good adhesion between matrix and reinforcements has been observed, and no debonding of particles during polishing has been detected. At higher-temperature processing, evaporation of magnesium was observed.

The presence of aluminum-magnesium spinel was found, which is probably formed by interaction of liquid magnesium with the aluminum oxide layer deposited on the ceramic particles during sintering. Thermodynamic calculations performed with the FactSage 5.0 database program [59] indicate the feasibility of formation of the surface oxide by direct oxidation of the AlN ceramics by traces of oxygen remaining in the nitrogen atmosphere; such a reaction shows a free energy formation of −864 kJ/mol at 1500 °C.

Chedru et al. [60] report the formation of amorphous alumina on the surface of AlN when infiltrating commercial aluminum alloys on particulate aluminum nitride preforms. In that case the preforms were preheated at 630 °C before being pressure

Fig. 4.14 SEM micrograph of Mg/AlN composite infiltrated at 930 °C [35]

infiltrated at 130 MPa. The spinel formation is attributed to the reaction between the magnesium element of the matrix and the Al_2O_3 formed on the surface of the ceramic AlN reinforcement following Eqs. (4.2), (4.3), and (4.4).

4.3.3   TEM Observations

Figure 4.15 shows a TEM bright-field image of Mg/AlN composite infiltrated at 900 °C. The ceramic reinforcement was immersed in the metal matrix, while $MgAl_2O_4$ spinel is present in the form of white lines in the select area of Fig. 4.15a. Usually this phase was found to precipitate close to the ceramic particles. Besides detecting the presence of the $MgAl_2O_4$ spinel at the interface, the formation of magnesium oxide (MgO) was also observed. Its formation is attributed to the reaction of the alumina film with magnesium according to Eq. (4.4), which thermodynamically is feasible.

4.3.4   Mechanical Characterization

The tensile properties of the composites were assessed by the shear punch test (SPT). Figure 4.16 shows the schematic representation of this setup and the obtainment of the specimens from the composite. The data generated from SPT have been

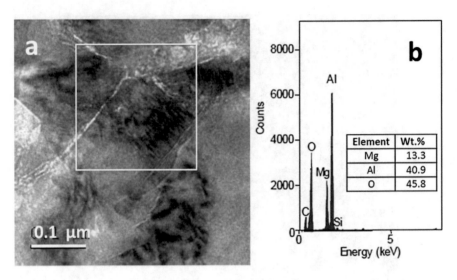

Fig. 4.15 (**a**) TEM image of Mg/AlN composite infiltrated at 900 °C showing the MgAl$_2$O$_4$ spinel, (**b**) EDX analysis of MgAl$_2$O$_4$ spinel

Fig. 4.16 Schematic of the small punch test setup

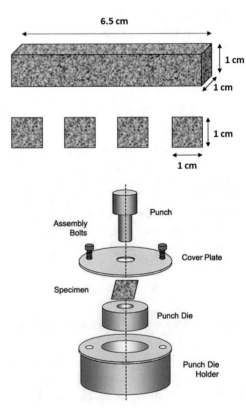

Table 4.7   Results of mechanical properties of Mg/AlN composites

Infiltration temperature (°C)	Hardness (HRB)	Elastic modulus (GPa)	UTS (MPa)[a]
870	82	108	319
900	79	106	313
930	68	87	278
960	77	94	280

[a]Ultimate tensile strength (UTS) was obtained as an average of four tests

processed as described in [58], and the ultimate tensile strength of Mg/AlN composites at various infiltration temperatures is given in Table 4.7. Four tests were carried out for each infiltration temperature, and an average of UTS is reported. In general, the lower the infiltration temperature, the higher the tensile strength. The UTS values obtained in this work are in the range reported in the literature of 304 MPa for Al/AlN composites containing 58 vol.% ceramic particles [60].

Additionally, elastic modulus (E) and HRB hardness measured on Mg/AlN composites are shown in Table 4.7. A decrease in modulus and hardness is clearly observed as the infiltration temperature increases because of a slightly higher porosity in the composites, which could be attributed to a higher evaporation of Mg. According to results reported by Lai and Chung [61], elastic modulus of Al/AlN composites containing 58 vol.% and 63 vol.% of AlN is 144.3 GPa and 163.5 GPa, respectively. It is clear that by increasing the volume fraction of the reinforcing phase, the elastic modulus of the composites increases.

Analysis of fracture after performing the SPT reveals an intergranular fracture type as is shown in Fig. 4.17. This implies that load transfer occurs between the matrix and reinforcements. Fracture through the particles is difficult because of their strong nature and small size.

4.3.5   Thermoelectric Characterization

Composites with a high content of reinforcement are essential to fit the thermal properties required for applications in the electronic field. The linear coefficient of thermal expansion (CTE) was $7.65 \times 10^{-6}/°C$ for Mg/AlN composites in the temperature range of 215–315 °C. This property is higher than the ceramic phase, which has a relatively low CTE ($4.4 \times 10^{-6}/°C$), since magnesium has a high CTE ($24 \times 10^{-6}/°C$). Moreover, the CTE of the Mg/AlN composites is lower than Al/AlN MMC's [60], which have a CTE of $9.81 \times 10^{-6}/°C$ in the temperature range of 35–100 °C and of $11.75 \times 10^{-6}/°C$ from 35 to 300 °C.

S-4700 5.0kV 4.5mm ×30.0k SE(U)   1.00um

Fig. 4.17   Fracture surface of Mg/AlN composite infiltrated at 900 °C [35]

4.4   Synthesis and Characterization of Al-Cu$_x$/TiC and Al-Mg$_x$/TiC Composites

Metal matrix composites consisting of Al-Cu$_x$ and Al-Mg$_x$ alloys reinforced with TiC particles have been fabricated using the pressureless infiltration method. Pressureless melt infiltration of ceramic preforms is an attractive technique for fabricating metal matrix composites. This process permits the formation of materials with a high ceramic content without the use of an external force. While aluminum alloys are the most common matrix employed in metal-ceramic composites, it is reported that the addition of TiC as a reinforcement improves both the mechanical properties and the strength at high temperature [62, 63]. In particular, the Al-Cu alloys/TiC system as a ceramic metal composite provides a favorable combination of electrical and mechanical properties [64, 65]. On the other hand, Al-Mg alloys/TiC are very light composites. Also, magnesium does not form stable carbides, so reinforcement carbides tend to be stable in pure magnesium [36].

Sintering and infiltrating temperatures play a very important role on the kinetics of infiltration. Therefore, a study on the infiltration rate is essential to determine the driving mechanism for a given infiltrating system. Previous studies have been focused on processing parameters such as porosity and infiltration temperature using commercial aluminum alloys [22, 36]. Some of these composites have been structurally and chemically characterized based on transmission electron microscopy (TEM) observations on as-fabricated and heat-treated composites [66, 67].

In this work, pressureless infiltration was used to produce metal matrix composites of Al-Mg and Al-Cu alloys reinforced with TiC particles [68]. The effect of Mg and Cu as alloying elements on the infiltration and kinetics behavior and some mechanical properties were investigated. The pressureless melt infiltration of Al-Cu and Al-Mg alloys into particulate 56 vol.% TiC preforms was studied. The infiltration of aluminum alloys varying the Mg and Cu content was compared with the infiltration rate of pure aluminum and was carried out in a thermogravimetric analyzer (TGA).

4.4.1 Experimental Conditions

TiC powders (H.C. Starck, grade c.a.s.) with an average particle size of 1.2 μm and a surface area of 2.32 m^2/g (Fig. 2.11) were used in the fabrication of porous preforms for liquid infiltration tests. The ceramic preforms were prepared by pressing uniaxially 18.5 g of powder at 8 MPa in a rectangular die to form green bars of ~6 × 1 × 1 cm in size. Sintering of the preforms was carried out in a tube furnace for 1 h under an argon atmosphere at 1250 °C. TiC preforms with 44% of porosity were fabricated for infiltration studies. Considering the dimensions of the bars before and after sintering, a shrinkage of 3.1% was obtained. The feasibility of sintering the TiC preforms was directly related to the fine particle size of the TiC powders.

The binary aluminum alloys were obtained by melting commercially pure aluminum in a graphite crucible with Cu or Mg additions. Pure copper and aluminum (99.99%) and electrolytic magnesium were used to prepare the binary alloys. The composition of the used alloys was 1, 4, 8, 20, and 33 wt.% for copper and 1, 4, 8, and 20 wt.% for magnesium. Pure commercial aluminum was also infiltrated for comparison purposes.

Al-Cu$_x$/TiC and Al-Mg$_x$/TiC composites were produced by pressureless melt infiltration of aluminum alloys into TiC preforms. Infiltration was carried out in a thermogravimetric analyzer (CAHN TG-2121) at 900 °C and 1000 °C under an argon atmosphere as is shown in Fig. 4.18. A detailed description was described in Sect. 3.1.1.1 and elsewhere [22]. The weight change in the preforms as a result of the infiltration of the liquid aluminum into the TiC preforms was monitored to obtain the characteristic infiltration profiles of the samples.

The resulting microstructures and reaction products on the composites were investigated by X-ray diffraction (XRD), scanning electron microscope (SEM), transmission electron microscopy (TEM), and energy-dispersive spectroscopy (EDS) analysis.

Additionally, Vickers hardness of the composites was determined using a standard Vickers tester under a 50 kg load, and the elastic modulus of the composites was measured using a GrindoSonic Lemmens equipment and the Halpin-Tsai [27] equation.

Fig. 4.18 Experimental setup for the kinetic infiltration

4.4.2 Infiltration Kinetics

Figure 4.19a shows the infiltration profiles obtained for the Al-Cu$_x$ binary alloys at 1000 °C. The infiltration rate of Al-Cu$_x$ increased with decreasing the copper content. A similar behavior was observed at 900 °C. This behavior is mainly due to the increased viscosity of the melt with copper additions, which in turn decreased the fluidity of the alloy through the capillaries. Contrary to the Al-Cu$_x$ system, the infiltration rate of Al-Mg alloys increased with increasing the Mg content (Fig. 4.19b).

The infiltration with Al-Mg alloys was made at 900 °C to prevent excessive evaporation of Mg. Considering that the infiltration rate is a function of the viscosity and surface tension of the liquid metal, Mg is capable of reducing both properties, leading to an improved motion of liquid into the porous network which in turn increases the fluidity of the molten metal [69]. The surface tension reduction is very sharp for the initial 1 wt.% Mg addition (from 860 mN/m to 650 mN/m). With further addition of Mg, the reduction is very marginal [70].

There is an incubation period present on the infiltration profiles that can be attributed to a transient contact angle between the liquid aluminum and TiC,

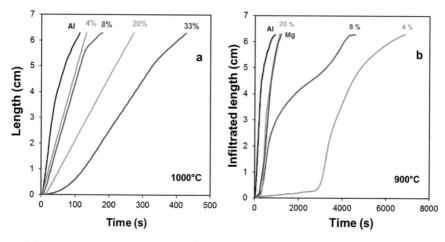

Fig. 4.19 (a) Infiltration profiles for Al-Cu$_x$/TiC composites and (b) for Al-Mg$_x$/TiC composites infiltrated at 900 °C [68]

influenced by the alloying elements. This incubation period also became less significant as the infiltration temperature increased. In the case of Al-Cu, it is reported that the surface tension of aluminum increases as the copper content increases [71]. However, increasing the Cu content decreases the melting point of the alloy from 660 °C for pure aluminum to 548 °C [72] at the eutectic composition (33 wt.% Cu).

For all the Al-Mg alloys and pure Mg, the infiltration rate never was greater than pure aluminum. Therefore, it is possible that the oxidation of Mg affects the infiltration process, which could be according to the reaction of Eqs. (4.1), (4.2), (4.3), (4.4), and (4.5).

If this reaction takes place, Mg tends to form a surface oxide in addition to the Al$_2$O$_3$ layer, inhibiting a real contact between the TiC and the aluminum. According to Lloyd [49, 73], high magnesium levels and low temperatures lead to MgO formation, while the spinel forms at very low Mg contents. In addition, it is believed that evaporation of Mg can help to disrupt the oxide layer of the melt to allow a real contact between TiC and aluminum. This phenomenon was present during infiltration, since powdered Mg was found on the inner surface of the tube after every experiment even though Mg has a boiling point near to 1120 °C, and the experiments were carried out in a dynamic atmosphere of argon. XRD analysis made to these powders revealed the formation of MgO. Because the oxidation of Al-Cu alloys is thermodynamically less feasible than oxidation of Al-Mg alloys, this could be a reason why they infiltrate faster than Al-Mg alloys.

Comparing both systems at 900 °C (Fig. 4.20), it is observed that the infiltration of TiC with Al-Cu alloys was faster than with Al-Mg in spite of the fact that the surface tension of the melt decreases with the Mg content. However, increasing the Mg content also increases the feasibility of MgO formation and the vaporization of Mg. There is an incubation period prior to stabilizing into parabolic-type infiltration

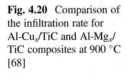

Fig. 4.20 Comparison of the infiltration rate for Al-Cu$_x$/TiC and Al-Mg$_x$/TiC composites at 900 °C [68]

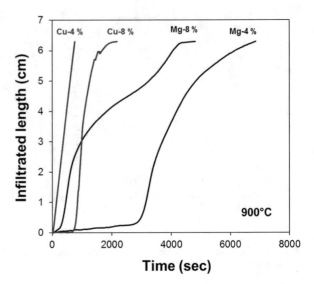

curves. This period is more noticeable for Al-Mg alloys and becomes less significant as the infiltration temperature increases. At low copper content, the infiltration was almost spontaneous. The decrease of the initiation time for the high Mg content samples could be attributed to the reduction of the surface tension of aluminum in addition to the reduction of the melting point. The more complex form of the curve at 8% Mg content is related to an increased evaporation of Mg as time elapses.

A kinetic study performed on the incubation periods for the Al/TiC systems that contain Mg and Cu as the principal alloying elements gave high activation energies (261 and 318 kJ/mol, respectively). These results suggested that infiltration process is driven by chemical reaction or a solid-state mechanism. Indeed, it is considered that the infiltration process is driven by a chemical reaction which becomes the rate-determining step in the infiltration process.

4.4.3   Microstructural Characterization of the Composites

Figure 4.21 shows the microstructure found in the Al-Cu$_x$/TiC samples infiltrated at 1000 °C; the dark color phase is the aluminum matrix, and the lighter phase surrounding the TiC grains is the CuAl$_2$ phase. Microstructural characterization carried out on the samples revealed the presence of CuAl$_2$ phase and TiC particles as the two most predominant phases distributed in the metallic matrix.

The presence of the CuAl$_2$ precipitates was confirmed by EDS and X-ray diffraction as is shown in Fig. 4.22.

The scanning electron microscopy observations indicated good interfacial bonding between the TiC particles and the Al-Cu matrix. Therefore, voids or other discontinuities were rarely observed at the particle-matrix interface. If the bonding

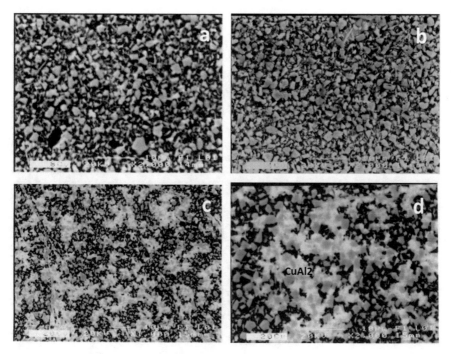

Fig. 4.21 Typical microstructure of the as-infiltrated (**a**) Al-1Cu/TiC, (**b**) Al-4Cu/TiC, (**c**) Al-8Cu/
TiC, and (**d**) Al-20Cu/TiC composites, infiltrated at 1000 °C [68]

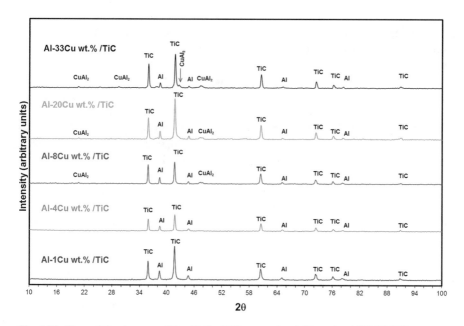

Fig. 4.22 X-ray diffractograms of the Al-Cu$_x$/TiC composites infiltrated at 1000 °C [68]

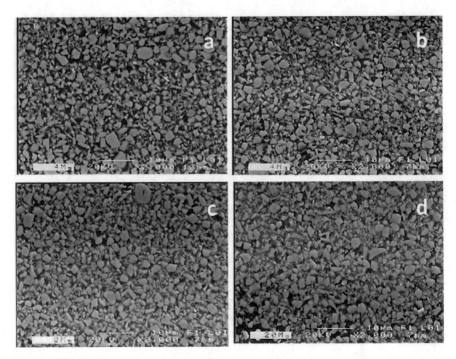

Fig. 4.23 Typical microstructure of the as-infiltrated (**a**) Al-1Mg/TiC, (**b**)Al-4Mg/TiC, (**c**)Al-8Mg/TiC, and (**d**)Al-20Mg/TiC composites, infiltrated at 900 °C

at the interface is stronger, the efficiency of load transfer to the ceramic reinforcement phase improves, and thereby, higher strength would be expected in the samples.

Figure 4.23 shows the typical microstructure of the as-infiltrated Al-Mg$_x$/TiC composites. For the composites processed with the Al-Mg alloys, no reaction phase was detected using XRD as is shown in Fig. 4.24. Total infiltration was achieved with Al-Mg$_x$ alloys, but no apparent interfacial reaction between the matrix and the ceramic was observed. It is observed from the micrographs that the earlier porous network was occupied by aluminum, forming a continuous interconnected metal matrix.

Magnesium does not form stable carbides, so carbides like TiC should be a suitable reinforcement for Mg matrix composites. However, if the Mg alloy contains elements, such as Al, which can react with carbides, then these carbides could not be fully stable.

A more complete study of the interfacial reactions that could be in the composite through transmission electron microscopy (TEM) was carried out. Figure 4.25 shows a TEM bright-field image of a specimen region from the Al-20Mg/TiC composite. This image shows the particles from the TiC uniformly distributed on the matrix (Al-20Mg). Chemical analysis from particles and matrix region was also obtained. This is illustrated in Fig. 4.25 (1 and 2), where a TiC stoichiometry and

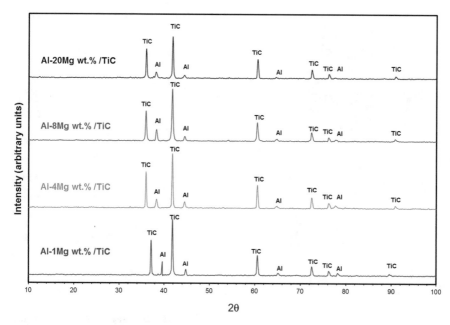

Fig. 4.24 X-ray diffractograms for the Al-Mg$_x$/TiC composites infiltrated at 900 °C [68]

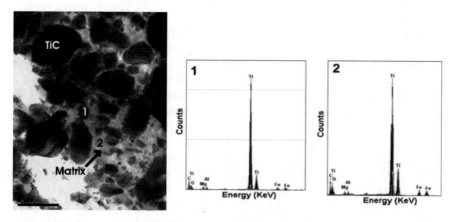

Fig. 4.25 TEM image (bright field) obtained from the Al-20Mg/TiC composite (*1*) and (*2*) X-ray compositional spectrum which shows the presence of Al, Mg, Ti, and Fe (the iron presence is related with the specimen-supporting grid) [68]

signals from Al and Mg can clearly be seen. The image shows precipitates of different sizes and morphologies (e.g., big particles of Ti are in the range of 2 μm and small particles until 90 nm).

Past investigations [51, 73, 74] on the magnesium additions have led to the spinel formation along the metal-ceramic interface. In this study, no spinel formation was detected with the XRD technique. However, TEM observations (Fig. 4.26) revealed

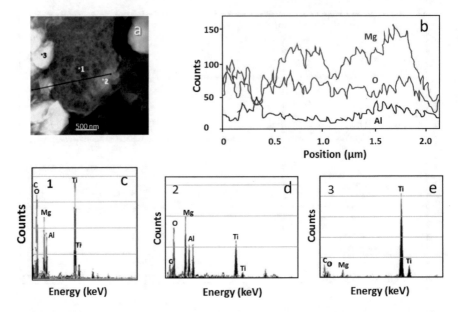

Fig. 4.26 (**a**) Dark-field image obtained from the Al-20Mg/TiC composite, (**b**) Line scan profiles performed in the path showed in Fig. 4.26a, (**c-d-e**) EDS spectrum in points 1, 2 and 3 showed in Fig. 4.26a [68]

localized spinel formation which gives rise to composites such as $MgAlO_2$ and Mg_2AlO_3. Figure 4.26a shows the dark field image obtained from the Al-20Mg/TiC composite, showing an agglomerate with some precipitates and some TiC particles. Figure 4.26b shows a line scan profiles performed in the Fig. 4.26a. Figure 4.26c-d-e show the EDS microanalysis at three points showed in Fig. 4.26a. Figure 4.26 (c) and (d) shows microanalysis results which suggest stoichiometries are closely related with the spinel formation. On the other hand, Fig. 4.26(e) shows the characteristic spectrum obtained from a TiC particle (white particle in Fig. 4.26a).

The X-ray line scan clearly shows the Al, O, and Mg decrements (Fig. 4.26a). When a TiC particle is detected, in this case a pronounce increment of the TiC signal is obtained (Fig. 4.26b).

The nature of the reaction between Al and TiC has not yet been definitively established. What is certain is that the two materials are not in equilibrium when they are in contact. Thermodynamic predictions suggest that TiC may be stable in liquid Al above a temperature of 752 °C, according to the reaction:

$$13Al_{(l)} + 3TiC_{(s)} = Al_4C_{3(s)} + 3TiAl_{3(s)}$$
$$\Delta G_{(750°C)} = -2.14\,kJ$$

(4.10)

Fine and Conley [75] discussed the Gibbs energy changes for the formation of TiC and Al_4C_3. They suggested that TiC shows a higher stability than Al_4C_3. Yokokawa et al. [76] obtained a chemical potential diagram for the Ti-C-Al system.

This diagram shows that Al, TiC, and Al3Ti at 1000 °C can coexist in equilibrium. Kennedy et al. [77] have indicated that in as-manufactured composites of Al/TiC, no reaction was observed; however, in the 48-h heat-treated (700 °C) composites, Al$_3$Ti and Al$_4$C$_3$ phases were obtained. Thermodynamic calculations by Frage et al. [78] indicate that a four-phase equilibrium between Al, TiC, Al$_3$Ti, and Al$_4$C$_3$ occurs at 693 °C and that only by heat treating below this temperature can both Al$_3$Ti and Al$_4$C$_3$ be produced. Even though the Al$_4$C$_3$ phase was not detected by XRD, the presence of this phase in the composites is feasible.

Comparing the Al-Cu$_x$/TiC and the Al-Mg$_x$/TiC composites, the systems containing Mg were at least ~5% lighter than the Al-Cu$_x$/TiC composites; the higher the Mg content, the lighter the composite.

The analysis made by SEM revealed a homogeneous distribution of CuAl$_2$ in the Al-Cu$_x$/TiC composites. The controlled precipitation and distribution of fine particles at room or elevated temperature are used to develop the mechanical properties of composites [67]. This copper phase is usually formed during the cooling process in the temperature interval of 520 °C to 500 °C [79].

4.4.4 Mechanical Characterization

The hardness and elastic modulus are shown in Table 4.8. The hardness of the Al-Cu$_x$/TiC and Al-Mg$_x$/TiC composites increased with increasing Mg and Cu content, and the amount of CuAl$_2$ precipitation increases. A maximum was reached in the eutectic composition, where the Al-33Cu/TiC composite achieved a hardness of 392 HV compared to 225 HV for the pure Al/TiC composite. In the case of the Al-Mg$_x$/TiC composites, the maximum hardness was 340 HV corresponding to the Al-20Mg matrix. Since the volumetric fraction in the ceramic preforms was the same in all the cases, the hardness values are a function of the matrix composition only.

For comparison, the theoretical modulus of the composites was also calculated using the Halpin-Tsai equation [27] due to its simplicity and semiempirical nature.

Table 4.8 Hardness and elastic modulus in function of the alloying element content

Composite	Hardness (HRV)	Elastic modulus (GPa)
Al-1Mg/TiC	262	170
Al-4Mg/TiC	285	164
Al-8Mg/TiC	315	160
Al-20Mg/TiC	340	150
Al-1Cu/TiC	257	172
Al-4Cu/TiC	263	174
Al-8Cu/TiC	291	187
Al-20Cu/TiC	354	195
Al-33Cu/TiC	392	180
Al/TiC	225	170
Mg/TiC	187	130

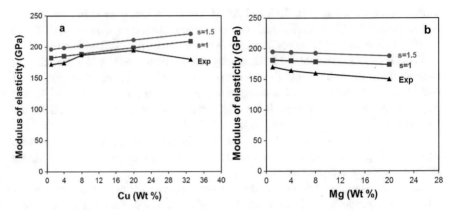

Fig. 4.27 Elastic modulus for composites in function of (**a**) Cu and (**b**) Mg content

The modulus was calculated for an aspect ratio of 1 and 1.5 as is shown in Fig. 4.27a and b. This equation is an approximation to predict the elastic modulus of composites reinforced with particles.

The results obtained with the Halpin-Tsai equation were found to be higher than the experimental values. These discrepancies in the results can be due to the presence of finite amount of porosity in the composites.

Figure 4.27 shows the effect of the alloying elements on the elastic modulus of the composites. It is important to mention that elastic modulus of the preforms before being infiltrated was in the order of 30–35 GPa. The elastic modulus of the Al-Cu$_x$/TiC samples has a tendency to increase with copper content. Meanwhile, the elastic modulus of the Al-Mg$_x$/TiC samples decreased slightly with the increase of Mg content in the alloys. Thus the elastic modulus of the pure Al/TiC composite (170 GPa) decreased by alloying with Mg. On the contrary, the elastic modulus of the Al-Cu$_x$/TiC composites increased by increasing the Cu content to reach a maximum of 195 GPa for the Al-20Cu matrix and then decreased slightly for the eutectic alloy.

4.5 Synthesis and Characterization of Al(1010-2024-6061-7075)/TiC Composites

MMCs reinforced with ceramic particles offer high-strength values and elastic modulus as well as good high-temperature properties compared with the corresponding monolithic materials used as the matrix. Aluminum is very attractive due to its low density, high thermal conductivity, and environmental resistance; however it has a high thermal expansion coefficient and low mechanical strength. The use of TiC as a reinforcement offers a high melting point and extremely high hardness, is relatively inexpensive, and exhibits good wetting characteristics by aluminum and its alloys. Therefore, TiC improves the aluminum properties mentioned above, as well as the resistance to high temperature.

4.5.1 Experimental Conditions

Commercial aluminum alloys (Al-1010, Al-2024, Al-6061, and Al-7075) were used to fabricate the composites reinforced with TiC by spontaneous infiltration process or pressureless infiltration. TiC porous preforms were obtained by sintering at 1250, 1350, and 1450 °C temperatures in argon atmosphere. Subsequently, these porous preforms were infiltrated by commercial aluminum alloys at different temperatures (950–1200 °C) [22]. The detailed processing of these compounds was described in Sect. 3.3. The chemical composition of each commercial alloy is shown in Table 4.9. It is important to mention that the 1010 aluminum alloy corresponds to aluminum of commercial purity (without alloying elements). This alloy will serve as a reference to compare the effect of the alloying elements.

One of the reasons for the use of these alloys was that it was thought that the alloying elements of these alloys would decrease the wetting angle with the TiC preforms, and therefore the rate of infiltration would be greater. However, the obtained results show that this is not the case; the alloying elements do not improve the wetting of aluminum by an interfacial chemical reaction compared to Al-1010 alloy [80, 81].

TiC wettability by Al-1010 alloy is better than the commercial Al alloys. Al-6061/TiC and Al-7075/TiC systems exhibit poor wetting. Wetting increased in the order 6061 < 7075 < 2024 < 1010 [82]. Figure 2.46 shows the wettability of these commercial alloys through sessile drop test at 900 °C temperature. The wettability increases with the reduction of the contact angle. The infiltration rates of these commercial alloys were slower than pure aluminum under similar conditions. However, some of these commercial alloys (Al-2024 and Al-6061) have the advantage that the produced composite can be heat-treated improving some mechanical properties [66, 67] and in some cases the corrosion resistance; such is the case of heat-treated composites Al-2024/TiC, Al-Mg$_x$/TiC, and Al-Cu$_x$/TiC, Ni/TiC, and Al-Cu-Li/TiC [83–86]. When a heat treatment of aging (either artificially or naturally) is applied to this composite type, the anodic corrosion current density increases, and the number and depth of pits decrease.

TiC powders had an average particle size of 1.25 μm with the distribution shown in Fig. 4.28a. Figure 4.28b shows a micrograph obtained by scanning electron microscopy (SEM) of TiC powder used. These powder features enhanced the powder sintering, obtaining porosities of 45, 41, and 36% in the preforms depending on the temperature.

Table 4.9 Chemical composition of the commercial aluminum alloys used in the fabrication of the composites

Alloy	Alloying element (wt.%)						
	Cu	Mg	Mn	Fe	Cr	Si	Zn
1010	0.005	0.014	0.002	0.65	0.012	0.20	0.002
2024	4.46	1.86	1.08	0.36	0.01	0.35	0.04
6061	0.25	0.82	0.06	0.40	0.18	0.61	0.07
7075	2.21	2.32	0.26	0.60	0.22	0.25	5.52

Fig. 4.28 (**a**) Particle size distribution of TiC powders used to produce the porous preforms; (**b**) SEM micrograph of TiC powders

Fig. 4.29 Micrographs obtained by SEM from TiC sintered preforms at (**a**) 1250 °C and (**b**) 1450 °C

4.5.2 Microstructural Characterization of the Composites

Figure 4.29 shows SEM micrographs of the partially sintered preforms at 1350 and 1450 °C. There is no significant difference between the sintering temperatures of 1350 and 1450 °C.

Once the preform sintered was obtained, the melt aluminum alloys were infiltrated, thus forming the composite material. Figure 4.30 shows SEM micrographs of the as-produced composites.

Infiltration was carried out in a thermogravimetric analyzer (CAHN TG-171) at the temperature range of 950–1200 °C under an argon atmosphere as is shown in Fig. 4.18. A full description of the method was described elsewhere [22]. The weight change in the preforms as a result of the infiltration of the liquid aluminum into the TiC preforms was monitored to obtain the characteristic infiltration profiles of the samples.

Fig. 4.30 Micrographs obtained by SEM from (**a**) Al-6061/TiC composites sintered at (**a**) 1250 °C and (**b**) 1450 °C, infiltrated at 1100 °C

The resulting microstructures and reaction products on the composites were investigated by X-ray diffraction (XRD), scanning electron microscope (SEM), transmission electron microscopy (TEM), and energy-dispersive spectroscopy (EDS) analysis.

4.5.3   Mechanical Characterization

In general, Al alloys/TiC composites have good physical and mechanical properties due to the high volume fraction of the reinforcement phase. The mechanical and physical properties of the composites produced depend mainly on the reinforcement content and the aluminum infiltration temperature. The elasticity modulus of the sintered preforms increases with the TiC preform density, that is, with the reinforcement content as is shown in Fig. 4.31.

It can be clearly seen that the elasticity modulus of the TiC sintered preforms increases noticeably with the addition of aluminum, more than three times. However, it is almost unaffected by the infiltration temperature as can be seen in Fig. 4.32.

The hardness increases with the increase of the sintering temperature, since it increases the density of the preform or reinforcement content (Fig. 4.33), but decreases with the increase in the infiltration temperature (Fig. 4.34).

This is attributed to the spheroidization of the TiC particles due to the apparent erosion of the interconnected necks of TiC grains by the high temperature of the molten aluminum, as can be seen in Fig. 4.30b.

Al-2024/TiC and Al-6061/TiC composites were heat-treated for improving their mechanical properties. Figure 4.35 shows SEM micrographs of Al-2024TiC composite as-fabricated (Fig. 4.35a) and heat-treated (Fig. 4.35b).

Table 4.10 shows the mechanical properties obtained from these composites. Mechanical properties of both heat-treated and as-fabricated composites showed a

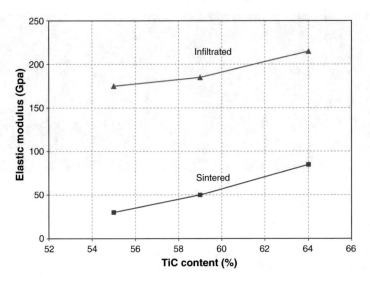

Fig. 4.31 Elasticity modulus for TiC preforms before infiltration and after being infiltrated with Al-6061 in function of TiC content

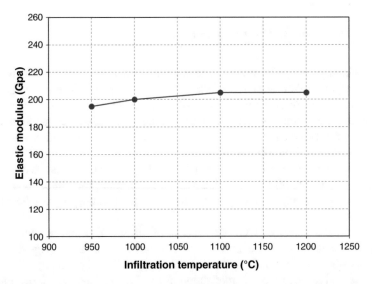

Fig. 4.32 Elasticity modulus for composites (Al-6061/TiC) in function of temperature for preforms sintered at 1450 °C

high dependence on the reinforcement content [22, 36, 66–68, 87]. Yield strength (YS), ultimate tensile strength (UTS), modulus of elasticity (E), and hardness (HRC) of the composite increase with an increase in the TiC volume content, whereas percent elongation decreases. The tensile properties increase significantly with the

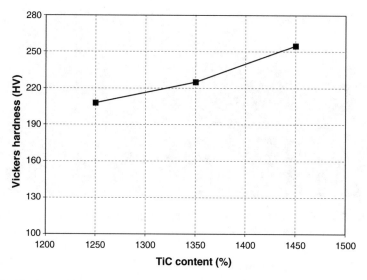

Fig. 4.33 Hardness of the sintered preforms at different temperatures

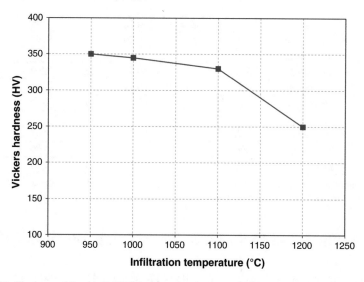

Fig. 4.34 Hardness of the Al-6061/TiC composites infiltrated at different temperatures (sintered at 1450 °C)

aging heat treatment for both Al-2024 and Al-6061/TiC composites. The highest strengths for Al-2024 and Al-6061/TiC composites were obtained with the solution and naturally aged heat treatment [67]. For the case of Al-6061/TiC, the fracture toughness decreases with the heat treatment, as can be seen in Table 4.10.

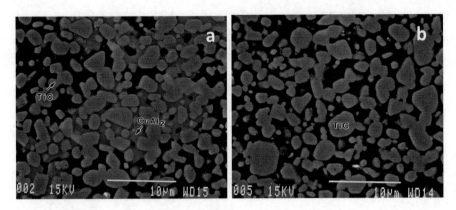

Fig. 4.35 SEM micrographs from Al-2024/TiC composite sintered at 1450 °C and infiltrated at 1200 °C, (**a**) as-fabricated and (**b**) heat-treated [67]

Table 4.10 Mechanical properties of the Al-2024 and Al-6061/TiC composites as-fabricated and heat-treated

TiC %	Condition	UTS (MPa)	YS (MPa)	EL (%)	HRC	E (GPa)	CVN (J)
Al-2024							
52	As-fabricated	360	205	0.28	26	195	–
	Artificially aged	400	260	0.20	36	197	–
	Naturally aged	465	298	0.17	35	197	–
55	As-fabricated	379	243	0.24	29	200	–
	Artificially aged	420	290	0.19	38	203	–
	Naturally aged	480	335	0.17	39	203	–
Al-6061							
52	As-fabricated	270	–	0.35	23	165	0.80
	Artificially aged	300	–	0.34	31	240	0.67
	Naturally aged	356	–	0.30	33	240	0.50

UTS ultimate tensile strength, *YS* yield strength, *EL* elongation, *HRC* hardness Rockwell C, *E* Young's modulus, *CVN* Charpy V-notch

4.6   Synthesis and Characterization of Composites with SiC

SiC is one ceramic of greater development used in structural applications and various engineering applications like in electrical and electronic industries, due to its excellent electrical, thermal, mechanical, and chemical properties. The properties of SiC are mainly due to its crystalline structure, which can be cubic or hexagonal, the latter being the most chemically and mechanically stable, even at temperatures above 2000 °C. Structure type obtained depends to a large extent on the method used for its production [88]. The high thermal conductivity, high melting point, relatively low thermal expansion, high specific resistance, excellent hardness, resistance to abrasion and corrosion, and mechanical stability at temperatures up to 1650 °C offer multiple opportunities for use [89], although one of the disadvantages is its fragility. Due to the properties mentioned above and to its abundance and low density, SiC is one of the reinforcements most widely used in MMC fabrication, combining their properties with the matrices used, which has generally low stiffness. In recent years SiC has been used in the MMC manufacture because they have good mechanical properties such as stiffness, wear resistance, high strength, and fatigue resistance. The manufacturing, aerospace, and civil and military industries are the main consumers of these composite materials [90].

Various techniques for the MMC processing have been developed for applications in the electrical, electronic, and aerospace industries. The manufacturing processes can be classified into two groups: liquid-state processing (infiltration, in situ deposition, and smelting) and solid-state processing [91]. Processes in the liquid state are more used than solid-state processes, because they are more economically and technically easier to process and fabricate, and some of these can be used on a large scale and with very varied shapes; the nature of the technique in the manufacture of composite materials has a marked effect on the properties of the material. In the case of processing in the liquid state for the manufacture of composite materials, the greatest difficulty that arises is the non-wettability of the ceramic phase by liquid metals. Therefore, it is of great importance to obtain good wettability to achieve a strong bond between the matrix and the reinforcement [14]. MMC's solid-state processing includes manufacturing by powder metallurgy and diffusion bonding. Chapter 3 describes most of the processes used in the MMC manufacture.

In recent years, studies have been conducted to evaluate the properties of MMCs manufactured by various techniques using SiC reinforcements. The infiltration of porous preforms of the reinforcement material with assisted pressure or pressureless infiltration and powder metallurgy are the most used techniques when a high amount of reinforcement (greater than 30% vol.) is required, mainly for the electronic industry, since these composites, in addition to good mechanical properties, achieve excellent thermal conductivities and low thermal expansion coefficients. In the case of compounds with amounts less than 30% reinforcement, methods such as disintegrated melt deposition or stir casting are used, this being one of the most economical manufacturing processes; however, they have the disadvantage that the dispersion of the reinforcement is not homogeneous, especially in amounts greater than 20%. These MMCs are being used with very good acceptance in the automotive industry, in aeronautics, and in general in several engineering applications due to their low density and good mechanical properties.

Chu et al. [92] studied the effect of porosity on the thermal conductivity (TC) of SiC_p/Al composites manufactured through sintering the spark plasma, concluding that thermal conductivity decrease with increasing porosity of the composite. These results were consistent with the approximate average effective model proposed in this investigation; also, they found that the larger the particle size of the SiC, the higher the TC of the composite. Chen et al. [93] investigated the thermal conductivity of a copper matrix composite reinforced with 55 vol.% SiC by pressure infiltration method and found that the TC increased slightly with the increase in particle size; however, the

TC of the composite is well below the one calculated theoretically by the Hasselman–Johnson [94] model due to the amount of high defects in the interface. One of the problems that occurs in the processing of compounds with SiC reinforcements and aluminum matrices or alloys containing this element is the formation of aluminum carbide, since some processes are carried out at high temperatures, which increases the metal-ceramic reactivity in the interface; therefore, it is important to take care of the interphase products that are generated between the matrix and the reinforcement. The possible interfacial reactions of the Al/SiC system have been studied in order to determine the products that occur in this system. The processes for the manufacture of MMC using the liquid phase are prone to the formation of aluminum carbide (Al_4C_3) according to Reaction (4.11):

$$3\,SiC_{(s)} + 4\,Al_{(l)} = Al_4C_{3(s)} + 3\,Si \tag{4.11}$$

Beffort et al. [95] who studied the mechanical properties of a compound with an aluminum matrix and 60 vol.% SiC elaborated by squeeze-casting method found that the MMC reaches values of modulus of elasticity of 200 GPa compared with 70 of aluminum; on the other hand, they report that Al_4C_3 only formed when magnesium was added to the matrix, which is justified by the simultaneous appearance of Mg_2Si, since the silicon coming from the dissolution of SiC in the liquid aluminum reacts instantaneously with the Mg according to Reactions (4.12) and (4.13):

$$2Mg + Si = Mg_2Si \tag{4.12}$$

$$4Al + 6Mg + 3SiC = Al_4C_3 + Mg_2Si \tag{4.13}$$

Likewise, they state that due to the speed of the squeeze-casting process, there are no kinetic conditions for the formation of Al_4C_3. Due to the excellent properties of the composites manufactured with SiC reinforcements and matrices with aluminum contents, it has been investigated on the reaction products between the matrix and the reinforcement, and recommendations have been made to inhibit the presence of undesirable reaction products such as aluminum carbide, among others. The addition of Si in the liquid matrix requires a minimum of 7 at.%, and 1 at.% in solid state [96] to avoid the formation of Al_4C_3. Ren et al. [97] carried out research on the effects of the addition of Si and Mg to the aluminum matrix in MMC manufactured by the pressureless infiltration route, finding that the addition of Si improves the thermal conductivity with additions of up to 6 wt.%, but with higher percentages (12%) decreases the thermal conductiviy; in the case of the addition of Mg to the matrix with 12 wt.% Si, the TC decreases drastically when more than 8 wt.% Mg is added and, in the same way, the density of the compound decreases. This behavior is attributed to the fact that with high amounts of magnesium, many pores are generated, since the magnesium evaporates at temperatures lower than that of aluminum; this also contributes to the formation of MgO within the pore; likewise, they report high values of modules of elasticity and bending strength with additions of 4 and 8 wt.% Mg, however 12 wt.% decreases the mechanical properties due to the porosity obtained at 12% Mg. To prevent interfacial reactions between SiC and Al several techniques has been studied [98, 99]. One of this techniques include the used of coating of SiC particles with metals such as copper or nickel, which improve wettability and reduce interfacial reactions, acting as a barrier between the matrix and the reinforcement [98].

Kim and Lee [99] studied the coating of SiC as a reinforcement of Al-2014 alloy; the MMC was manufactured by the route of hot-pressed in vacuum and later extrusion. The SiC particles were oxidized at the temperature of 1100 °C for 1, 2, 4, and 6 h, found at the interface as reaction product the spinel $MgAl_2O_4$, which has better mechanical and thermal properties than aluminum carbide and silicon dioxide, also reported the Si, as reaction product instead of Al_4C_3, the best value of interfacial bonding strength was presented at 1 h oxidation of the reinforcement. It is important to take care of the exposure time of the carbide to the oxidizing atmosphere. Xue and Yu [100] concluded that short exposure of silicon carbide does not prevent the formation of aluminum carbide because the silicon dioxide layer does not completely coat the particles. An adequate exposure time causes the entire surface to be coated, and this, in turn, reacts totally with the alloy forming the magnesium spinel. On the other hand, a prolonged exposure to the oxidizing environment generates a layer that is too thick, so that one part reacts with the aluminum, magnesium, and oxygen present, and another part of the dioxide remains intact. Silicon dioxide, while not as harmful as aluminum carbide, also decreases the thermomechanical properties of the compound. Regarding the bending strength, the best value was reported when the reinforcement was maintained for 4 h and a thermal treatment T6 was applied to the composite.

4.6.1 Mg-AZ91E/SiC Composites

Zalapa [101] studied the thermal conductivity and coefficient of thermal expansion of a MMC made with an AZ91E magnesium alloy reinforced with SiC particles coated with silicon dioxide, the MMC was manufactured by pressureless infiltration route, and the particles used had an average size of 7.25 and 21 μm. The oxidation of the porous SiC preforms was carried out in a tubular furnace with an air flow at a temperature of 1200 °C for 2 h. The formation of the SiO_2 layer induced by the oxidizing atmosphere can be seen in Fig. 4.36a. The fracture of the layer was induced by polishing process. The EDS chemical maps corroborate that there is undeniably formation of assumed layer. As can be seen in Fig. 4.36a, the fracture of the SiO_2 layer is observed, and in Fig. 4.36b a more intense tone can be seen revealing the presence of SiC without the layer that was removed with the polishing process.

Similarly, the presence of silicon dioxide was confirmed by the X-ray technique [102]. The porous, pre-sintered, and oxidized preform was infiltrated at temperatures between 750 and 900 °C under a flow of argon and nitrogen in three stages; in the first stage, argon was used until reaching the programmed temperature for

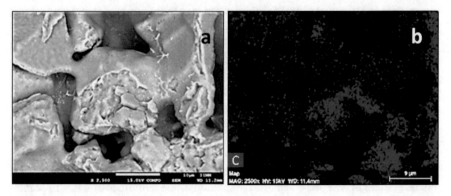

Fig. 4.36 SEM micrographs of oxidize SiC showing (**a**) layer of SiO_2 and (**b**) EDS of carbon

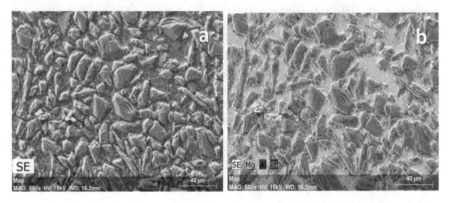

Fig. 4.37 SEM micrograph of (**a**) Mg-AZ91E/SiC composite and (**b**) elemental chemical mapping

infiltration; in the second stage, nitrogen was used for 15 min to promote infiltration; and in the third stage, argon was used for cooling process. The best values of density, hardness, and modulus of elasticity were obtained at the infiltration temperature of 750 °C since at higher temperatures there was greater porosity due to the magnesium vapor entering the ceramic preform generating a positive pressure and preventing or retarding the infiltration of the matrix [97].

Figure 4.37 shows the SEM micrograph of the composite Mg-AZ91E/SiC in which a homogeneous distribution of the reinforcement in the matrix can be seen and little porosity of the infiltrated MMC at 750 °C.

It was found that most of the ceramic surface was polished without detachment of ceramic particles, which reveals that adhesion between matrix and reinforcement is strong enough to support the preparation of the sample. It is presumed that this good adhesion is due to the formation of aluminum and magnesium spinels, the product of the reaction between SiO_2, and the liquid matrix containing Al and Mg [97].

The formation of the spinel at the interface occurs according to Reaction (4.14), since at the temperature of 750 °C (1023 K), there are thermodynamic conditions for the formation of the spinel or the reaction in solid state between magnesium oxide and the alumina present according to Reaction (4.15):

$$Mg + 2Al + 2SiO_2 \rightarrow MgAl_2O_4 + 2Si$$
$$\Delta G_{(750°C)} = -421.435 \, kJ \tag{4.14}$$

$$MgO_{(S)} + Al_2O_{3(S)} = MgAl_2O_{4(S)}$$
$$\Delta G_{(1000°C)} = -49.4 \, kJ/mol \tag{4.15}$$

4.6.2   Thermomechanical Properties

The density of the compounds was evaluated by the Archimedes method described by ASTM C20-00 [48]. Densities of 98.89, 97.42, and 82.06 g/cm^3 were obtained at temperatures of 750, 800, 850, and 900 °C, respectively. The decrease in density is due to the increase in porosity promoted by the higher generation of Mg vapor that occurs at a higher temperature, generating a positive pressure which prevents or delays the infiltration of the matrix [97]. Some thermomechanical properties of Mg-AZ91E/SiC composites were measured and they were compared with thoeretical calculations.

The average value measured of the modulus of elasticity of the Mg-AZ91E/SiC composites was 140 GPa for the composites with reinforcements of 21 μm as the average size and 160 GPa for MMC reinforced with particles 7.5 μm; the experimental values are close to those calculated by the Halpin-Tsai model [27] using the factor (aspect ratio) equals to 0.8, 1, and 2. An increase of elastic modulus of 318 and 367% is observed with respect to the matrix with the reinforcements of 21 μm and 7.5 μm, respectively. The hardness obtained for composites infiltrated at 750 °C was 286 HV and 318 HV with reinforcements of 21 and 7.5 μm of SiC, respectively. An increase in hardness is observed with the decrease in particle size, and increases of 427% and 475% are reached for particle sizes of 21 μm and 7.5 μm, respectively, with respect to the matrix. The composites produced achieved a CTE of 10.7 × 10^{-6} °C^{-1} for composites made with the finest particles and 11.5 × 10^{-6} °C^{-1} for those made with thicker particles in the temperature ranges between 25 and 300 °C. The values obtained in this study are close to those calculated by the Kerner model [39, 103]. The results of thermal conductivity (TC) obtained for the composites with SiC reinforcements with two different particle sizes have a marked difference since the TC at 25 °C for the composite manufactured with particle sizes of 21 μm is 92 W/m K and 46 W/m K for the composite made with particles of average size of 7.5 μm. The difference in thermal conductivities is attributed to the increased in connectivity of the composite manufactured with coarse particles with respect to which was synthesized with fine particles, because this promotes a larger amount of thermal barriers and results in greater dispersion of the matrix electrons, which are responsible for heat conduction. The TC decreases with the increase in temperature reaching values of up to 54 W/m K for coarser particle reinforcement and 24 W/m K for the finest particle reinforcement at the temperature of 300 °C. The decrease in TC is attributed to the decrease in the mean free path of phonons and electrons. When the temperature increases, in the case of phonons, the harmonic movement of the atoms becomes inharmonic, which prevent them from spreading through the material. In the same way, it happens with the electrons, which accelerate to a point where they interfere with their own movement.

MMC manufactured with SiC reinforcements that amounts to less than 30 vol.% uses routes such as hot-pressing, metallurgy power, or stir casting processes. Casting smelting processes have been a favored processing method, as they lend themselves to the manufacture of the large number of components of complex shapes. Especially, stir casting route is mostly used to produce MMCs reinforced with particles, because it has shown promise for the manufacture of MMC forms in a simple, effective, and low-cost manner. Thus, a large amount of MMC manufactured with percentages lower than 30% have been studied with SiC reinforcements in low-density matrices, which achieve very good properties of high strength to weight ratio, hardness, and wear resistance. Basavarajappa et al. [104] studied the wear resistance of an MMC manufactured by liquid metallurgy with an Al-Cu-Mg alloy reinforced with SiC_p (15 vol.%) plus 3% graphite; they report an increase in wear resistance compared to the alloy especially when the sliding speed is high. In the same way, Prakash et al. [105] investigated a compound made with a pure magnesium matrix with reinforcements of 5 and 10 wt.% SiC with additions of graphite by the powder metallurgy route and found that microhardness, density, and wear resistance increased with the content of SiC. By the method of stir casting, an Al-Si-Mg/10%vol. SiC composite was manufactured, finding that the temperature and the holding time reduce the viscosity and that the ultimate strength increases between 700 and 800 °C but then decreases; also this property decreases with holding time [106].

4.6.3   Mg/SiC Composites

Arreola [107] made an Mg/SiC composite by stir casting route. The elaboration of the MMC was carried out in a simple device that consisted of a carbon steel crucible with a mold and a particle adder. A vertical steel crucible and an electrical engine double blade stirrer with progressive angle from 90° to 60° with a range of 280–2200 rpm and Ar (99.95%) were used (Fig. 4.38). The incorporation of the SiC particles was performed at 750 °C by stirring during 5 min and subsequent casting of the melt into a preheated mold. The experimental setup was designed in a single device to maintain the inert atmosphere at all time.

Figure 4.39 shows the SEM micrograph of the Mg/20 vol.% SiC composite in which a good distribution of the SiC particles is observed in the matrix. It is also observed that there are apparently no reaction products at the interface, although

Fig. 4.38 The experimental
setup of crucible and mold
devise

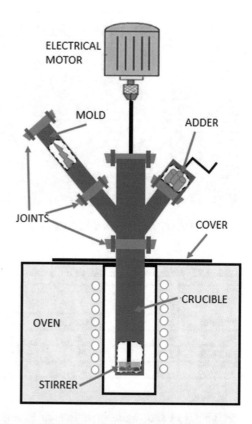

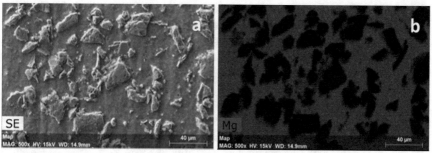

Fig. 4.39 SEM micrographs of Mg/SiC 20 vol.%, (**a**) secondary electrons and (**b**) elemental
mapping of Mg

thermodynamically it is possible that the formation of reaction products like MgO
and Si, by means of X-ray diffraction and EDS, was not identified.

Figure 4.40 presents optical micrographs of Mg/SiC$_p$ composites with 10% and
20 vol.% SiC, manufactured at 750 °C, where a good distribution is appreciated, but
in the case of the composite with 20% reinforcement, more porosity and the
tendency to form clusters are observed.

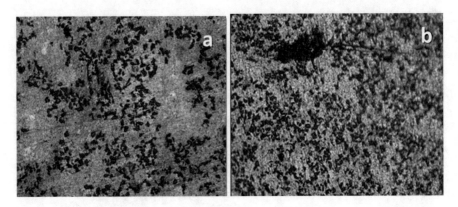

Fig. 4.40 Optical micrographs of SiC/Mg composites (**a**) 10 vol.% SiC and (**b**) 20 vol.% SiC

Table 4.11 Mechanical properties of the composites in function of SiC vol.%

Material	Density (g/cm^3)		Young's modulus (GPa)		Hardness (HV)
	Mixture rule	Experimental	Halpin-Tsai	Experimental	
Mg	–	1.78	45	45	27.9
SiC 10%	1.88	1.83	52.7	49.3	49.1
SiC 20%	2.06	1.83	61.7	53.4	45.2

Table 4.11 shows the values attained for the composites with reinforcements of 10 and 20 vol.% SiC of reinforcements compared with the Mg matrix and with the values calculated by mathematical models. It can be seen that in the case of density, the MMC manufactured with 10% reinforcement reaches values very close to those calculated by the rule of the mixtures, while those made with 20% reinforcement have a greater deviation, resulting in lower density.

Regarding the Young's modulus, it is observed that the values obtained by the manufactured composites are close to those calculated by the Halpin-Tsai equation [27]. It can be seen that the Young's modulus of the composites manufactured with a 10% reinforcement has an increase of around 11%, while the one manufactured with 20% reinforcement of SiC reaches an increase of less than 19%, which is attributed to the amount of pores that are generated in this MMC by increasing the amount of reinforcement. The values in the hardness tests were very consistent in all the points where the indentation was carried out, so it is assumed that there was a good distribution of the particles in the magnesium matrix; the hardness of the compound with 10% reinforcement increased 75% with regard to the matrix; however, by increasing the percentage of reinforcement up to 20% with an average result of 45.26 HV, the hardness in the composite was not as expected, since a greater amount of reinforcement caused a greater porosity.

4.7 Synthesis and Characterization of Composites with Al$_2$O$_3$

The possibility of some improvement of mechanical properties of oxide ceramics by manufacturing of particulate composites has been very well recognized recently. Among the oxide ceramics, zirconia and alumina are the most important materials, widely used in structural applications, due to their properties. Fabrication of two-phase particulate composites could be the simplest way to the improvement of mechanical properties. Despite a wide range of alumina-zirconia composites, non-oxide particles were also often utilized as strengthening agents [108]. Ceramic materials used in femoral heads for total hip replacement have to combine good mechanical and tribological properties, associated with perfect in vivo stability. Alumina and yttrium-stabilized zirconia (Y-TZP) are considered as premium choices today [109].

As a member of one of the most common groups of advanced ceramics, alumina possesses great potential to be used at high temperatures because of its excellent resistance to heat, corrosion, wear, and oxidation. Alternatively, its application has yet been restricted due to its low fracture toughness. If cracks and flows acting as fracture initiation are completely repaired, the reliability of structural ceramics will be greatly improved, and alumina ceramics would be a leading candidate material for advanced gas turbine and engine and gas-tight tubes [110]. In general, one way to overcome this problem is to enhance the fracture toughness of the ceramic matrix reinforcing it by fibers and whiskers and ductile or brittle particles [111, 112]. Most of the work on microscale composites has been concerned with alumina matrices. Alumina-metal nanocomposites can be produced by hot-pressing powder blends of either alumina and metal powders or alumina and metal oxide powders. Although composites containing relatively high amounts of nanoscale metallic particles have been reported, in general only a relatively small addition is required to produce a significant improvement in strength.

4.7.1 Ni/Al$_2$O$_3$ Composites

Pulsed electric current sintering (PECS), sometimes called plasma-activated sintering (PAS) or spark plasma sintering (SPS), process is a newly developed method for achieving sintering with minimum grain growth for a short time [113, 114]. Due to the high heating efficiency, PECS is a process that can easily consolidate a high-quality specimen at lower sintering temperature in a shorter time than conventional sintering methods, mainly by means of a spark pulse current directly into the powder particles and mold [115, 116]. As such, PECS is a very effective process for the sintering of advanced materials such as ceramics, intermetallic composites, nanocrystalline materials, and functionally gradient materials, which are difficult to sinter by the conventional sintering method. Basically, raw powder in a graphite die is pressed uniaxially, and pulsed direct current

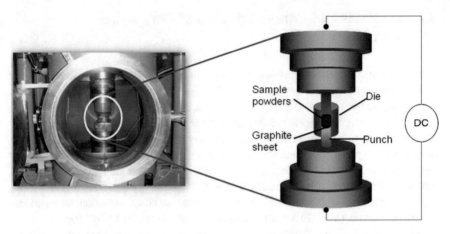

Fig. 4.41 Schematic representation of the sample setting inside the sintering equipment [120]

(DC) momentarily is applied through the electrically conducting pressure die and, in appropriate cases, also through the sample to heat itself [117, 118]. Several ceramics such as Si_3N_4, AlN, Al_2O_3, and SiC have successfully been sintered by this method [113–118]. Figure 4.41 shows a schematic representation of the sample setting inside the sintering equipment.

4.7.2 Experimental Conditions

Salas-Villaseñor et al. [119, 120] studied the characterization of alumina (Al_2O_3) composites toughened by Ni-dispersed nanoparticles produced by pulsed electric current sintering (PECS).

Al_2O_3-based nanocomposites (CMC) with 5 vol.% Ni particle dispersion was fabricated in four steps: (a) slurry preparation, (b) drying, (c) powder reduction, and (d) electric current sintering. This process is illustrated in Fig. 4.42.

For the preparation of Ni/Al_2O_3, powder mixtures, commercial α-alumina powder (Sumitomo Chemical Co., AA-04, with mean grain size 0.4 μm and purity 99.99%), and Ni-nitrate (Ni $(NO_3)_2 \cdot 6H_2O$, purity 99.9%, mean grain size of 200 nm) as the source material for Ni particles were selected. Ni/Al_2O_3 nanocomposite was prepared by an aqueous solution method; the content of Ni in the final composite was fixed to be 5 vol.%. The slurry of α-Al_2O_3 powder and Ni-nitrate in water was dried at 300 °C and then reduced by raising temperature to 600 °C with a heating rate of 400 K/h for 12 h in a stream of Ar + 1%H_2 gas mixture. The nanocomposite powder was sintered to a disc (15 mm diameter × 4 mm thickness) using pulsed electric current sintering (PECS) at 1400 °C for 5 min in vacuum under uniaxial pressure of 45 MPa.

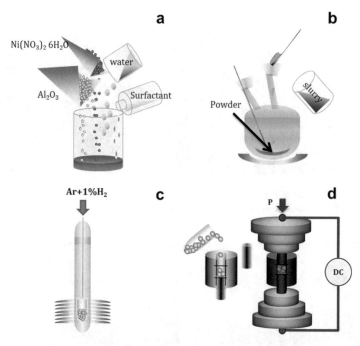

Fig. 4.42 Preparation of the Ni/Al$_2$O$_3$ ceramic metal composite (CMC): (**a**) slurry preparation, (**b**) drying, (**c**) powder reduction, and (**d**) electric current sintering [119, 120]

4.7.3   Microstructural Characterization of the Composites

The relative density of the as-sintered Ni/Al$_2$O$_3$ samples (4.02 $\pm$ 0.04 g/cm^3) was measured by the Archimedes technique using toluene, attained at least 98% of the theoretical value (4.15 g/cm^3). This value for the relative density of the samples produced by PECS is similar or lower to that one obtained for samples produced by hot-pressing (98%) [121] or by slip casting under traditional sintering without pressing (99.72%) [122], respectively; however lower temperature and time are required by PECS process. On the other hand, the results for relative density of Ni/Al$_2$O$_3$ composite materials produced by PECS are higher than relative density of the materials produced by infiltration methods [112]. Figure 4.43a shows the nanocomposite powder, and Fig. 4.43b shows the as-sintered Ni/Al$_2$O$_3$ nanocomposite.

Ni particles of less than 100 nm and 0.5 µm in sizes can be appreciated before sintering process. The sintered disc was fractured, and microstructural observation of the surface was performed by scanning electron microscopy (SEM). Figure 4.44 shows the fracture surface of the Ni/Al$_2$O$_3$ nanocomposite. Ni particles of less than 100 nm and 0.5 µm in sizes can be appreciated before (Fig. 4.43) and after sintering process. It is possible to observe in Fig. 4.44 a homogeneous distribution of the spherical Ni particles that takes place in the grain boundaries of the Al$_2$O$_3$ matrix, as has been reported by different authors [123, 124].

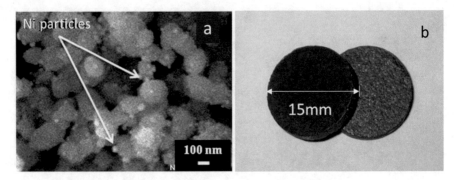

Fig. 4.43 SEM micrograph of (**a**) nanocomposite powder and (**b**) original Ni/Al₂O₃ nanocomposites sintered at 1400 °C for 5 min using PECS [120]

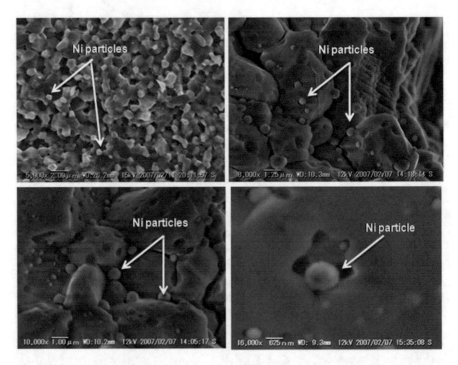

Fig. 4.44 SEM micrographs of fracture surface of Ni/Al₂O₃ nanocomposites after sintering at 1400 °C for 5 min using PECS [120]

4.7.4 Mechanical Characterization

The fracture toughness of the sintered material was evaluated by JIS R 1607 (Japanese Industrial Standard related to Fine Ceramics) method using Miyoshi's equation (4.16) [125, 126]:

Fig. 4.45 Schematic representation during fracture toughness evaluation of Ni/Al$_2$O$_3$ composites

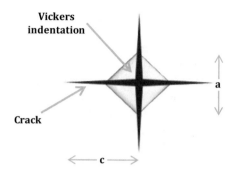

$$K_{1c} = 0.026 \frac{E^{0.5}(P)^{0.5}a}{c^{1.5}} \tag{4.16}$$

where K_{1c} represents the fracture toughness (MPa m$^{\frac{1}{2}}$), E is the Young's modulus (N/m^2), P is the load at fracture (49.03 N), a represents the dimensions of indentation (m), and c is the length of the crack measured starting of the center of indentation (m). A value of 380 GPa was used for the Young's modulus for all the samples [127]. Figure 4.45 shows a schematic representation of the system to measure the different variables to evaluate the fracture toughness of Ni/Al$_2$O$_3$ composites.

The average fracture toughness values of the three Vickers indentations by sample and of K_{IC} for the data evaluated is 7.0 ± 0.14 MPa m$^{\frac{1}{2}}$. This value is higher than that of monolithic alumina (3–4 MPa m$^{\frac{1}{2}}$) [127] composites produced by infiltration (4.2 MPa m$^{\frac{1}{2}}$) [123] and even than the value reported for composites produced by hot-pressing process (3.5–4.2 MPa m$^{\frac{1}{2}}$) [122].

4.8 Synthesis and Characterization of Mg-AZ91E/TiC Composites

In the last years, Mg and TiC are the two components mostly used in the fabrication of metal matrix composites, due to its mechanical properties that can be obtained, easy processing, and lightweight compounds, among others.

Metal matrix composites (MMCs) have been extensively studied in the last years; one of the most important characteristics is the ability to tailor its properties, such as mechanical and electrical properties through the appropriate control of reinforcement and matrix. In addition, the different processes allow the fabrication of complicated shaped parts.

Aluminum was one of the most used materials in the fabrication of MMC; however, in the last years, magnesium and its alloys have been extensively used as a matrix. Magnesium and its alloys offer low melting point, good dimensional stability, and high thermal conductivity and do not react with TiC. AZ91E magnesium alloy is the most common magnesium casting alloy. It has a good combination with castability, mechanical strength, and good ductility and has excellent corrosion resistance. TiC is a hard ceramic material with a low thermal expansion coefficient, and it is thermodynamically stable, which combined with Mg permits to improve mechanical properties of the composites [1, 7, 8, 11, 13, 16]. The infiltration technique is one of the most used to fabricate Mg/TiC composites, because it permits to add high content of the ceramic reinforcement and allows to obtain isotropic properties in the composites.

In this study Mg-AZ91E/TiC$_p$ composite were fabricated using a spontaneous infiltration of Mg alloy at 950 °C [128, 129]. The composites produced have 37 vol. % of metal matrix and 63 vol.% of TiC-like reinforcement. The obtained composites were subsequently solution heat-treated at 413 °C during 24 h, cold water quenched, and subsequently artificially aged at 168 °C and 216 °C during 16 h in an argon atmosphere. The effect of heat treatment on the microstructure and mechanical properties was evaluated.

The influence of the solubilization heat treatment and aging hardening of Mg-AZ91/TiC composites was studied. The effect of aging behavior has been evaluated using Vickers hardness and elastic modulus measurement analysis, related with microstructure developed during heat treatment. Time of aging was correlated with measures of elastic modulus and hardness.

4.8.1　Experimental Conditions

Mg-AZ91E/TiC composites were fabricated using the infiltration technique without external pressure. Powders of TiC with an average particle size of 7.4 μm and commercial AZ91E magnesium alloy were used to fabricate the MMCs. Figure 4.46 shows a SEM image of the TiC powders and distribution of particle size. Figure 4.47 shows a SEM image of the Mg-AZ91E alloy used. Chemical composition of the magnesium alloy is shown in Table 4.12. Some of the most important mechanical and physical properties of Mg alloy and TiC are shown in Table 4.13.

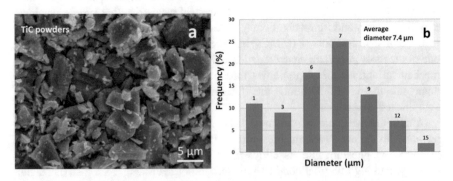

Fig. 4.46 (a) SEM image of TiC powders and (b) distribution of particle size [129]

Fig. 4.47 SEM image of Mg-AZ91E alloy used in the fabrication of composites [129]

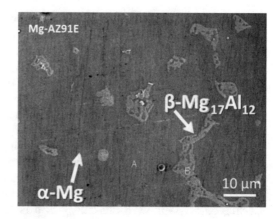

Table 4.12 Chemical composition of the Mg-AZ91E alloy (wt.%)

Al	Zn	Mn	Si	Fe	Cu	Ni	Mg
8.80	0.71	0.19	0.029	0.001	0.002	0.001	Bal.

Table 4.13 Properties of AZ91E magnesium alloy and TiC

Property	AZ91	TiC
Elastic modulus (GPa)	45	450
Tensile strength (MPa)	117	258
Hardness (HB)	70	121
CTE (μm/m K)	27×10^{-6}	7.9×10^{-6}
Thermal conductivity (W/m K)	72	17.2
Density (g/cm^3)	1.8	4.93
Melting point ($^\circ$C)	470	3160

The preforms were fabricated by cold pressing TiC powders in a steel mold, with a load of 75 MPa in order to obtain green preforms of 6.5 × 1 × 1 cm. Green preforms were sintered at 1250 °C during 1 h under argon atmosphere in order to avoid oxidation of preforms. Porosity of sintered preforms as well as the composite was evaluated using the Archimedes method described in ASTM C20-00 [48]. Then, sintered preforms were infiltrated with AZ91E magnesium alloy at 950 °C.

After obtaining Mg-AZ91E/TiC$_p$ composites, a T6 heat treatment was applied. This treatment consists of two stages: solubilization and artificial aging. In the first stage, solubilization consisted of heating the composite at 413 °C for 24 h, in order to dissolve precipitates, after composites were cold water quenched. In the second stage, the composite was artificially aged, by heating the composite at 168 and 216 °C during 16 h in an argon atmosphere. A solubilization heat treatment was used to homogenize the matrix and decrease the Mg$_{17}$Al$_{12}$ precipitates [130].

An insight of the mechanical properties measuring hardness and elastic modulus of the composites was obtained. The hardness of the materials was determined through a Vickers NANOVETA indenter using 20 kg of load. At least three bars were used, and each bar has five indentations to obtain an average of the results.

Evaluation of the modulus of elasticity using the GrindoSonic MK5 JV Lemmens equipment was carried out, in which three bars of the composite were prepared to obtain a representative average; measurements in each of the four faces were performed. In addition, the modulus of elasticity was evaluated with a theoretical model using the equation of the mixture rule, the Hashin-Shtrikman and Halpin-Tsai models [26, 27].

Cross section of the composites with and without heat treatment was analyzed by X-ray diffraction technique using a SIEMENS D5000 diffractometer from 20° to 90° 2 theta and microstructurally characterized using scanning electron microscopy (SEM) JEOL 6400 coupled with an energy-dispersive spectroscopic (EDS). The reaction products in the interface were analyzed using transmission electron microscopy (Philips Tecnai F20) with 200 kV of acceleration voltage and equipped with a spectroscopic system of X-ray energy dispersive to perform microanalysis. The samples observed using TEM and high resolution (HRTEM) were prepared using small foils (300 μm) obtained from the composite materials, following a procedure of conventional preparation. Samples of 3 mm of diameter were thinning using a Dimpler D500i equipment; later the samples were electropolished using a Struers Tenupol-5 equipment with a 10% perchloric acid electrolyte, 40 V, and temperature of −10 °C for a time of 15 min.

4.8.2   Microstructural Characterization of the Composites

Figure 4.48 shows a SEM micrographs of the Mg-AZ91E/TiC composite material as-fabricated and with heat treatment, in which it is observed a homogeneous distribution of reinforcement material on the metal matrix. The composites produced have 37 vol.% of metal matrix and 63 vol.% of reinforcement (TiC).

The bulk density of the Mg-AZ91E/TiC composites was evaluated according to Archimedes method giving densities around 98.3% and around 1.7% of close porosity. Several elemental chemical analysis tests were carried out in different samples of the composite in order to show the elemental distribution. Figure 4.49

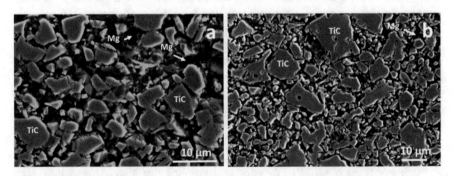

Fig. 4.48 SEM microstructures for Mg-AZ91E/TiC composites, (**a**) as-fabricated, (**b**) with solution heat treatment at 413 °C for 24 h and aged at 168 °C for 16 h

Fig. 4.49 SEM micrograph
of Mg-AZ91E/TiC
composite as-fabricated and
elemental chemical mapping
[129]

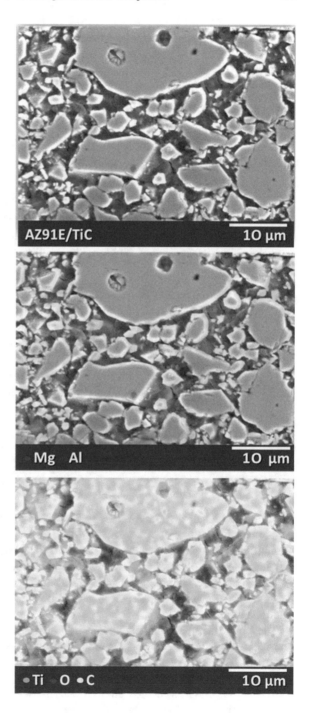

Fig. 4.50 X-ray diffraction patterns for Mg-AZ91E/TiC composites, (**a**) as-fabricated, (**b**) with solution heat treatment at 168 °C for 16 h

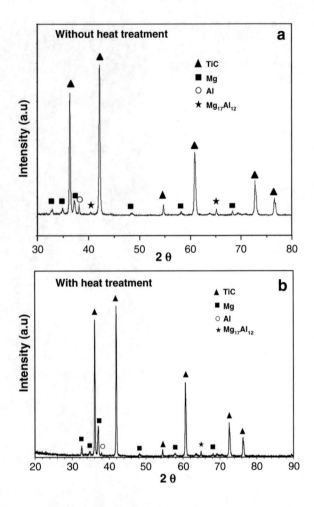

shows an elemental distribution mapping of the main elements of the composite as-fabricated.

In this figure some precipitates can be observed. Apparently the blue zones could be alumina or $MgAl_2O_4$ spinel. Even if the sintering and infiltration were carried out in argon, oxygen is not completely eliminated. It is possible that oxygen came from superficially oxidized preforms (TiO_2), and oxidizing it is possible after sintering the preforms during handling.

The composite materials without and with heat treatment were microstructurally characterized using X-ray diffraction; diffractograms obtained are shown in Fig. 4.50. Through X-ray diffraction, TiC, Mg, Al, and $Al_{12}Mg_{17}$ phases in the as-fabricated and heat-treated composites could be identified.

The composites heat-treated showed a decrease in the intensity of $Al_{12}Mg_{17}$ phase, which means that solubilization heat treatment met the homogenization of the matrix and decreased the $Mg_{17}Al_{12}$ precipitates.

TiC is thermodynamically stable in pure Mg on the system Mg/TiC [8, 131], but it can reacts with the microalloying elements of MgAZ91E forming interfacial products. However, any evidence of these reaction products was observed in the composite interface.

On the other hand, magnesium has high reactivity with oxygen to form MgO. It is possible that the formation of MgO from the reaction between Mg and TiO_2 may form during the sintering processes of ceramic preforms of TiC. Also MgO can be formed from the reduction of Al_2O_3 due to magnesium oxide that is thermodynamically more stable than alumina.

Studies on thermodynamic stability of Al-Mg oxides in Al-Mg alloys [49, 51] show that the formation of Al_2O_3, $MgAl_2O_4$, and MgO is possible. Reyes et al. [129] show a thermodynamic diagram of O_2 vs. composite composition (wt.%) for AZ91E/TiC system, showing the feasible reactions that may take place after applying T6 heat treatment. This graph predicts the appearance of different phases at temperature of 168 °C; mainly the formation of MgO can be observed and the presence of the $Mg_{17}Al_{12}$ phase, followed by the $MgAl_2O_4$ spinel.

Figure 4.51 shows a mapping carried out to the composite, focusing on the region of some precipitates in the matrix. Al, O, and Mg were detected in this region. According to the thermodynamic study, formation of $MgAl_2O_4$ spinel could be possible. The presence of an aluminum-magnesium spinel precipitate ($MgAl_2O_4$) was detected on the heat-treated composites. This spinel could be formed by

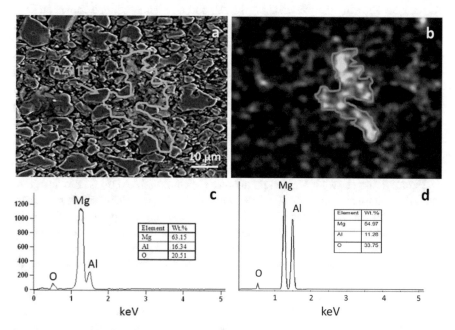

Fig. 4.51 (**a**) SEM micrograph of AZ91E/TiC composite heat-treated at 168 °C for 16 h and (**b**) elemental chemical mapping showing the region with elemental composition of Mg, Al, and O, (**c, d**) results of the mapping showing the stoichiometry of $MgAl_2O_4$ spinel [129]

interaction of liquid magnesium with aluminum oxide during the infiltration process (at 950 °C) or by solid-state reaction of magnesium oxide with aluminum oxide during the heat treatment (at 413 °C) of the composite according to Reactions (4.17) and (4.18):

$$3Mg_{(l)} + 4Al_2O_{3(s)} \rightarrow 2Al + 3MgAl_2O_4$$
$$\Delta G_{(950°C)} = -208.523\,kJ/mol \tag{4.17}$$
$$\Delta G_{(413°C)} = -204.668\,kJ/mol$$

$$MgO_{(s)} + Al_2O_{3(s)} \rightarrow MgAl_2O_4$$
$$\Delta G_{(950°C)} = -31.132\,kJ/mol \tag{4.18}$$
$$\Delta G_{(413°C)} = -27.626\,kJ/mol$$

4.8.3   TEM Observations

In order to analyze the interfacial reactions between the matrix and reinforcement, high-resolution transmission electron microscopy (HRTEM) was used. Figure 4.52a shows a high-angle annular dark-field (HAADF) STEM image of a precipitate in the interface between TiC particles and Mg matrix. Figure 4.52b shows a linear elemental analysis showing the presence of Mg, Al, and O. Figure 4.52c, d shows an EDS performed in two points of this precipitate that revealed the presence of Mg, Al, O, Ti, and C, which could be forming $MgAl_2O_4$ spinel.

After careful indexing, the HRTEM images show the reaction products that were identified as nanoparticles of spinel ($MgAl_2O_4$), magnesium oxide (MgO), and alumina (Al_2O_3). Figure 4.53 shows a HRTEM image of $MgAl_2O_4$ spinel identified by this technique. Figure 4.53a shows one small nanoparticle, identified as $MgAl_2O_4$ with an interplanar distance of 0.243 nm, which corresponds to the (311) plane distances. Figure 4.53b shows the FFT spectrum and filtered HRTEM images of the small nanoparticle region, and Fig. 4.53c shows the EDS spectrum of the particle showed in Fig. 4.53a.

Figure 4.54a shows a high-angle annular dark-field (HAADF) STEM image of another precipitate in the interface between TiC particles and Mg matrix. Figure 4.54b shows a linear elemental analysis showing the presence of Mg, Ti, and O. Figure 4.54c shows a linear elemental analysis showing the presence of Mg and O which could form magnesium oxide (MgO). Figure 4.54d shows a linear elemental analysis showing the presence of Ti and C which correspond to a TiC particle.

The analysis of the fine precipitates was analyzed by TEM. Figure 4.55a shows a precipitate with interplanar distances of 0.243 nm corresponding to MgO. And the EDS analysis suggests the presence of the magnesium and oxygen. Figure 4.55b shows the FFT obtained from HRTEM image of the MgO precipitate; this precipitate

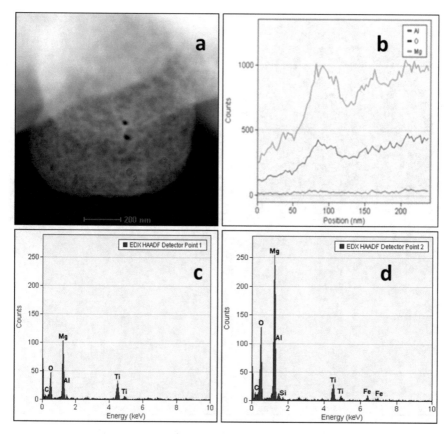

Fig. 4.52 (**a**) High-angle annular dark-field (HAADF) STEM image of a precipitate, (**b**) linear elemental analysis showing the presence of Al, O, and Mg, and (**c, d**) EDS performed in two points of this precipitate [129]

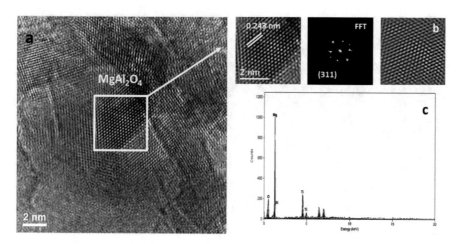

Fig. 4.53 (**a**) HRTEM image showing the presence of MgAl$_2$O$_4$ precipitates showing the interplanar distance of 0.243 nm corresponding to (311) plane and FFT obtained from HRTEM image, (**b**) power spectrum and filtered HRTEM images of the small nanoparticle region, (**c**) EDS spectrum of the particle [128]

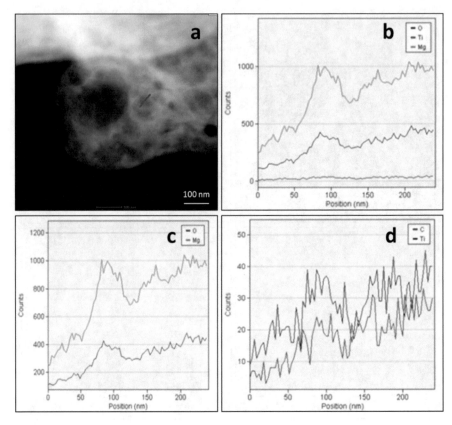

Fig. 4.54 (**a**) High-angle annular dark-field (HAADF) STEM image of a precipitate in the interface between TiC particles and Mg matrix, (**b–d**) linear elemental analysis showing the presence of Mg, O, Ti, and C [129]

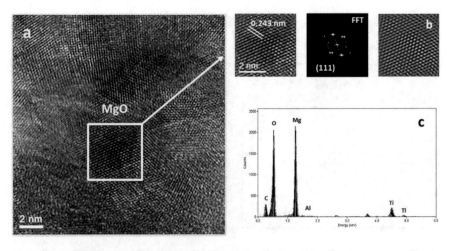

Fig. 4.55 (**a**) HRTEM image of the MgO phase showing the interplanar distance corresponding to (111) plane and FFT obtained from HRTEM image, (**b**) power spectrum and filtered HRTEM images of the small nanoparticle region, (**c**) EDS spectrum of the particle [128]

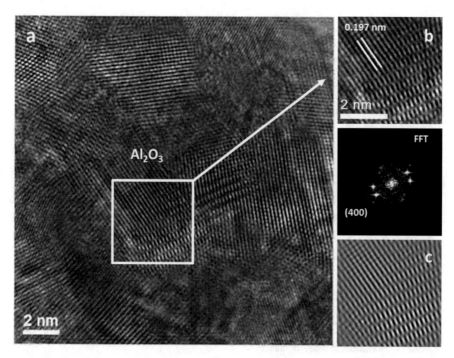

Fig. 4.56 (**a**) HRTEM image of AZ91/TiC composite heat-treated showing the presence of Al$_2$O$_3$ particle, (**b**) small nanoparticle region with interplanar distance of 0.197 nm corresponding to the distances between the (400) atomic planes, (**c**) power spectrum and filtered HRTEM images of the small nanoparticle region [128]

is oriented along the [1 −1 0] zone axis. Figure 4.55c shows the EDS spectrum of the particle showed in Fig. 4.55a.

Figure 4.56a shows a HRTEM image of the heat-treated AZ91E/TiC composites at 413 °C for 24 h and aged at 168 °C for 16 h. This image is related with an alumina showing some nanoparticles identified as Al$_2$O$_3$ with interplanar distances of 0.197 nm, which corresponds to the distances between the (400) planes. On the other hand, Fig. 4.56b shows the fast Fourier transform (FFT) spectrum obtained from the HRTEM image of the Al$_2$O$_3$ nanoparticle. It seems to be a cubic diffraction pattern. Figure 4.56c shows the spectrum and filtered HRTEM images of the small nanoparticle region.

4.8.4 Mechanical Characterization

Some mechanical properties of Mg-AZ91E/TiC composite were evaluated, such as hardness and elastic modulus, which directly depend on reinforcement content.

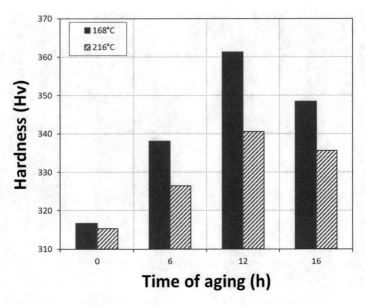

Fig. 4.57 Effect of aging temperature and time on the hardness

4.8.4.1   Hardness Measurements

The hardness of the composites was determined through a Vickers indenter. Three random bars were used for this purpose, and each bar has five indentations to obtain an average of the results. The hardness of the composites before applying a heat treatment has an average value of 316 HV (33 HRC). The hardness of the composites after carrying out the solution heat treatment and aging at 168 °C reaches a value of around 362 HV. The effect of heat treatment and time of aging at 168 and 216 °C is shown in Fig. 4.57.

4.8.4.2   Elastic Modulus

The elastic modulus was measured through GrindoSonic equipment. Three bars of the composite were used to obtain a representative average. The average value of elastic modulus for the as-fabricated composite was 162 GPa. The elastic modulus for these composites is similar to the elastic modulus reported by similar systems elsewhere [19, 42, 131]. However the volumetric reinforcement and particle size are different; in addition the elastic modulus of aluminum is higher than AZ91 alloy. The effect of heat treatment and time of aging at 168 and 216 °C in the elastic modulus is shown in Fig. 4.58.

Theoretical evaluations of elastic modulus were carried out by means of mixture rule, which calculates the elastic modulus in function of the volumetric fractions. In similar way using the Halpin-Tsai [27, 132] equation, the elastic modulus was

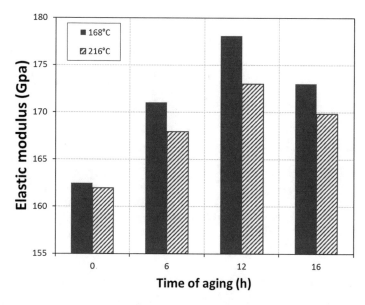

Fig. 4.58 Effect of aging temperature and time on the elastic modulus

Table 4.14 Modulus of elasticity of the composite measured and calculated by equations

Elastic modulus	Measured	Halpin-Tsai [27, 132] $S = 1$	Hashin-Shtrikman [26, 43]	Anasori [19]
E (GPa)	162	165	140	174

evaluated. This equation predicts in a better way the elastic modulus and takes into account the elastic modulus of each individual phase and the aspect ratio of the particles. Finally, using the Hashin-Shtrikman equation [26, 43], the elastic modulus was evaluated. Table 4.14 shows some results of these evaluations and comparison with measured values.

Halpin-Tsai model is a mathematical model for the prediction of elasticity of composite material based on the geometry (aspect ratio) of the reinforcement and the elastic properties of the reinforcement and matrix; thus better predictions are obtained.

4.8.4.3   Effect of Heat Treatment

The mechanical response of AZ91E/TiC composites was evaluated after applying a solution heat treatment at 413 °C for 24 h and aging at 168 and 216 °C up to 16 h. Aging is a process that can occur at room temperature (natural aging) or at temperatures established (artificial aging), where in a controlled manner, it produced

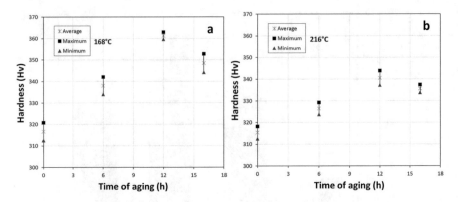

Fig. 4.59 Effect of the aging time on the hardness after solubilization heat treatment at 413 °C and aging at (**a**) 168 °C and (**b**) 216 °C

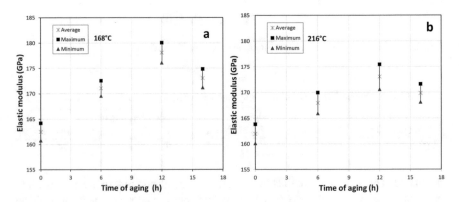

Fig. 4.60 Effect of the aging time on the elastic modulus after solubilization heat treatment at 413 °C and aging at (**a**) 168 °C and (**b**) 216 °C

formation of precipitates that is well distributed. This process is responsible to give the required final strength, modulus, and hardness.

Hardness and elastic modulus were evaluated in order to determine the effect of solubilization heat treatment at two different aging temperatures. Figures 4.59 and 4.60 show the hardness and elastic modulus behavior in function of the aging time, respectively. As can be observed, the elastic modulus was improved with the heat treatment (Fig. 4.60), obtaining better results with aging at 168 °C. The highest elastic modulus was obtained after 12 h of aging. Increasing the aging time beyond 12 h, the elastic modulus decreases, which may be attributed to an overage. A similar behavior was observed for the hardness (Fig. 4.59).

During the aging, fine precipitates are developed in the grains with homogenous distribution reaching a maximum peak in the mechanical properties [128]. Using long periods of time during aging, these precipitates grow too much, and the mechanical properties decrease producing an overage. Xiuqing et al. [133] studied the effect of aging time on the hardness of TiC/AZ91 composite at 175 and 200 °C. The composite reached peak hardness after 30 h at 175 °C and 12 h at 200 °C.

In summary, $AZ91E/TiC_p$ composite was fabricated by the pressureless infiltration method, after the effect of heat treatment on the microstructure and mechanical properties was evaluated. Mg, TiC, Al, and $Mg_{17}Al_{12}$ phases through XRD were detected. Meanwhile, using TEM analysis in heat-treated composites, $MgAl_2O_4$, MgO, and Al_2O_3 were identified. The as-fabricated composites have elastic modulus and hardness of 162 GPa and 316 HV, respectively. After solution heat treatment and aging at 168 °C during 12 h, the composites reach values of 178 GPa and 362 HV for the elastic modulus and hardness, respectively. Results revealed that increasing the aging time up to 12 h, the mechanical properties reach the maximum. After this time an overage may be produced, and mechanical properties decrease.

References

1. Dey A, Pandey KM (2015) Magnesium metal matrix composites – a review. Rev Adv Mater Sci 42:58–67
2. Luo A (1995) Processing, microstructure, and mechanical behavior of cast magnesium metal matrix composites. Metall Mater Trans A 26:2445–2455
3. Lopez VH, Truelove S, Kennedy AR (2003) Fabrication of Al–TiC master composites and their dispersion in Al, Cu and Mg melts. Mater Sci Technol 19:925–930
4. Sun XF, Wang CJ, Deng KK, Kang JW, Bai Y, Nie K, Shang SJ (2017) Aging behavior of AZ91 matrix influenced by 5 µm SiC_p: investigation on the microstructure and mechanical properties. J Alloys Compd 727:1263–1272
5. Wang XJ, Xu L, Hu XS, Nie KB, Deng KK, Wu K, Zheng M (2011) Influences of extrusion parameters on microstructure and mechanical properties of particulate reinforced magnesium matrix composites. Mater Sci Eng A Struct Mater 528:6387–6392
6. Shen MJ, Ying T, Chen FY, Hou JM (2017) Microstructural analysis and mechanical properties of the AZ31B matrix cast composites containing micron SiC particles. Int J Met Cast 11(2):287–293
7. Chen L, Yao Y (2014) Processing, microstructures, and mechanical properties of magnesium matrix composites: a review. Acta Metall Sin 27:762–774
8. Contreras A, Lopez VH, Bedolla E (2004) Mg/TiC composites manufactured by pressureless melt infiltration. Scr Mater 51:249–253
9. Dong Q, Chen LQ, Zhao MJ, Bi J (2004) Synthesis of TiC_p reinforced magnesium matrix composites by in situ reactive infiltration process. Mater Lett 58:920–926
10. Cao W, Zhang C, Fan T, Zhang D (2008) In situ synthesis and damping capacities of TiC reinforced magnesium matrix composites. Mater Sci Eng A 496:242–246
11. Jo I, Jeon S, Lee E, Cho S, Lee H (2015) Phase formation and interfacial phenomena of the in-situ combustion reaction of Al-Ti-C in TiC/Mg composites. Mater Trans 56:661–664
12. Chen L, Guo J, Yu B, Ma Z (2007) Compressive creep behavior of TiC/AZ91D magnesium-matrix composites with interpenetrating networks. J Mater Sci Technol 23(02):207–212
13. Lim CYH, Leo DK, Ang JJS, Gupta M (2005) Wear of magnesium composites reinforced with nanosized alumina particulates. Wear 259:620–625
14. Contreras A, Leon CA, Drew RAL, Bedolla E (2003) Wettability and spreading kinetics of Al and Mg on TiC. Scr Mater 48:1625–1630

15. Xiuqing Z, Haowei W, Lihua L, Naiheng M (2007) In situ synthesis method and damping characterization of magnesium matrix composites. Compos Sci Technol 67:720–727
16. Jiang QC, Li XL, Wang HY (2003) Fabrication of TiC particulate reinforced magnesium matrix composites. Scr Mater 48:713–717
17. Balakrishnan M, Dinaharan I, Palanivel R, Sivaprakasam R (2015) Synthesize of AZ31/TiC magnesium matrix composites using friction stir processing. J Magnes Alloys 3:76–78. https://doi.org/10.1016/j.jma.2014.12.007
18. Gu XY, Sun DQ, Liu L (2008) Transient liquid phase bonding of TiC reinforced magnesium metal matrix composites (TiCP/AZ91D) using aluminum interlayer. Mater Sci Eng A 487:86–92
19. Anasori B, Caspi N, Barsoum MW (2014) Fabrication and mechanical properties of pressure-less melt infiltrated magnesium alloy composites reinforced with TiC and Ti_2AlC particles. Mater Sci Eng A 618:511–522
20. Kaneda H, Choh T (1997) Fabrication of particulate reinforced magnesium composites by applying a spontaneous infiltration phenomenon. J Mater Sci 32:47–56
21. Ye HZ, Liu XY (2004) Review of recent studies in magnesium matrix composites. J Mater Sci 39:6153–6171
22. Contreras A, Salazar M, León CA, Drew RAL, Bedolla E (2000) The kinetic study of the infiltration of aluminum alloys into TiC. Mater Manuf Process 15(2):163–182
23. Muscat D, Drew RAL (1994) Modeling the infiltration kinetics of molten aluminum into porous titanium carbide. Metall Mater Trans 25A(11):2357–2370
24. Massalski TB (ed) (1990) Binary alloy phase diagrams, vol 3, 2nd edn. American Society for Metals, Metals Park
25. Shimada S, Kozeki M (1992) Oxidation of TiC at low temperatures. J Mater Sci 27:1869
26. Hashin Z, Shtrikman S (1962) On some variational principles in anisotropic and non-homogeneous elasticity. J Mech Phys Solids 10:335–342
27. Halpin-Tsai JC (1992) Primer on composite materials analysis, 2nd edn. Technomic, Lancaster, pp 165–191
28. Boccaccini AR, Fan Z (1997) A new approach for the Young's modulus-porosity correlation of ceramic materials. Ceram Int 23:239–245
29. Elsayed A, Kondoh K, Imai H, Umeda J (2010) Microstructure and mechanical properties of hot extruded Mg–Al–Mn–Ca alloy produced by rapid solidification powder metallurgy. Mater Des 31:2444–2453
30. Tian J, Shobu K (2004) Hot-pressed AlN–Cu metal matrix composites and their thermal properties. J Mater Sci 39:1309–1313
31. Ye HZ, Liu XY, Luan B (2005) In situ synthesis of AlN in Mg–Al alloy by liquid nitridation. J Mater Process Technol 166:79–85
32. Mirshahi F, Meratian M (2012) High temperature tensile properties of modified Mg/Mg_2Si in situ composite. Mater Des 33:557–562
33. Huang Z, Yu S, Liu J, Zhu X (2011) Microstructure and mechanical properties of in situ $Mg_2Si/AZ91D$ composites through incorporating fly ash cenospheres. Mater Des 32:4714–4719
34. Swaminathan S, Srinivasa RB, Jayaram V (2002) The production of AlN-rich matrix composites by the reactive infiltration of Al alloys in nitrogen. Acta Mater 50:3093–30104
35. León CA, Arrollo Y, Bedolla E, Drew RAL (2006) Properties of AlN-based magnesium-matrix composite produced by pressureless infiltration. Mater Sci Forum 502:105–110
36. Contreras A, López VH, León CA, Drew RAL, Bedolla E (2001) The relation between wetting and infiltration behavior in the Al-1010/TiC and Al-2024/TiC systems. Adv Technol Mater Mater Process 3(1):33–40
37. Xiu Z, Yang W, Chen G, Jiang L, Ma K, Wu G (2012) Microstructure and tensile properties of Si_3N_4p/Al-2024 composite fabricated by pressure infiltration method. Mater Des 33:350–355
38. Ding-Fwu L, Jow-Lay H, Shao-Ting C (2002) The mechanical properties of AlN/Al composite fabricated by squeeze casting. J Eur Ceram Soc 22:253–261

39. Zhang Q, Chen G, Wu G, Xiu Z, Luan B (2003) Property characteristics of AlN/Al composite fabricated by squeeze casting technology. Mater Lett 57:1453–1458
40. Goh CS, Soh KS, Oon PH, Chua BW (2010) Effect of squeeze casting parameters on the mechanical properties of AZ91-Ca Mg alloys. Mater Des 31(suppl. 1):S50–S53
41. Chedru M, Vicens J, Chermant L, Mordike BL (1999) Aluminium–aluminium nitride composites fabricated by melt infiltration under pressure. J Microsc 196:103–112
42. Contreras A, Angeles-Chavez C, Flores O, Perez R (2007) Structural, morphological and interfacial characterization of Al-Mg/TiC composites. Mater Charact 58(8–9):685–693
43. Couturier R, Ducret D, Merle P, Disson JP, Jouvert P (1997) Elaboration and characterization of metal matrix composite: Al/AlN. J Eur Ceram Soc 17:1861–1866
44. Lai SW, Chung DD (1994) Fabrication of particulate aluminum matrix composites by liquid metal infiltration. J Mater Sci 29(12):3128–3150
45. Taheri-Nassaj E, Kobashi M, Chou T (1995) Fabrication of an AlN particulate aluminum matrix by a melt stirring method. Scr Mater 32:1923–1927
46. Wang L, Zhang BP, Shinohara T (2010) Corrosion behavior of AZ91 magnesium alloy in dilute NaCl solutions. Mater Des 31(2):857–863
47. Bedolla E, Lemus-Ruiz J, Contreras A (2012) Synthesis and characterization of Mg-AZ91/AlN composites. Mater Des 38:91–98
48. ASTM C20–00 (2000) Standard test method for apparent porosity, water absorption, apparent specific gravity and bulk density by boiling water. American Society for Testing and Materials
49. Lloyd DJ (1994) Particle reinforcement aluminum and magnesium matrix composites. Int Mater Rev 39:1–23
50. McLeod AD, Gabryel CM (1992) Kinetics of growth of spinel $MgAl_2O_4$ on alumina particulate in aluminum alloys containing magnesium. Metall Mater Trans 23A:1279–1283
51. Lloyd DJ, Lagacé HP, McLeod AD (1990) Interfacial phenomena in metal matrix composites. In: Ishida H (ed) Controlled interfaces in composites materials. Elsevier Science, New York
52. Contreras A, Bedolla E, Pérez R (2004) Interfacial phenomena in wettability of TiC by Al–Mg alloys. Acta Mater 52:985–994
53. Zheng M, Wu K, Yao C (2001) Characterization of interfacial reaction in squeeze cast SiC_w/Mg composites. Mater Lett 47:118–124
54. Zheng MY, Wu K, Kamado S, Kojima Y (2003) Aging behavior of squeeze cast SiC_w/AZ91 magnesium matrix composite. Mater Sci Eng A 348:67–75
55. Taheri-Nassaj E, Kobashi M, Choh T (1995) Fabrication of an AlN particulate aluminium matrix composite by a melt stirring method. Scr Mater 32:1923–1929
56. Chedru M, Boitier G, Vicens J, Chermant JL, Mordike BL (1997) Al/AlN composites elaborated by squeeze casting. Key Eng Mater 132–136:1006–1009
57. Baik Y, Drew RAL (1996) Aluminum nitride: processing and applications. Key Eng Mater 122–124:553–570
58. León CA, Drew RAL (2002) Small punch testing for assessing the tensile strength of gradient Al-Ni/SiC composites. Mater Lett 56:812–816
59. FactSage 5.0, Bale CW, Pelton AD, Thompson WT. Ecole Polytechnique de Montréal/Royal Military College, Canada (http://www.crct.polymtl.ca)
60. Chedru M, Vicens J, Chermant JL, Mordike BL (2001) Transmission electron microscopy studies of squeeze cast Al–AlN composites. J Microsc 201:299–315
61. Lai SW, Chung DDL (1994) Superior high-temperature resistance of aluminium nitride particle-reinforced aluminium compared to silicon carbide or alumina particle-reinforced aluminium. J Mater Sci 29:6181–6198
62. Kennedy AR, Wyatt SM (2000) The effect of processing on the mechanical properties and interfacial strength of aluminum/TiC MMC's. Compos Sci Technol 60:307–314
63. Muscat D, Shanker K, Drew RAL (1992) Al/TiC composites produced by melt infiltration. Mater Sci Technol 8(11):971–976
64. Frage N, Froumin N, Dariel MP (2002) Wetting of TiC by non-reactive liquid metals. Acta Mater 50(2):237–245

65. Rambo CR, Travitzky N, Zimmermann K, Greil P (2005) Synthesis of TiC/Ti–Cu composites by pressureless reactive infiltration of TiCu alloy into carbon preforms fabricated by 3D-printing. Mater Lett 59:1028–1031

66. Albiter A, Contreras A, Bedolla E, Pérez R (2003) Structural and chemical characterization of precipitates in Al-2024/TiC composites. Compos Part A 34:17–24

67. Albiter A, León CA, Drew RAL, Bedolla E (2000) Microstructure and heat-treatment response of Al-2024/TiC composites. Mater Sci Eng A289(1):109–115

68. Contreras A, Albiter A, Bedolla E, Perez R (2004) Processing and characterization of Al-cu and Al-Mg base composites reinforced with TiC. Adv Eng Mater 6(9):767–775

69. Goicoechea J, García-Cordovilla C, Louis E, Pamies A (1992) Surface tension of binary and ternary aluminum alloys of the systems Al-Si-Mg and Al-Zn-Mg. J Mater Sci 27:5247–5252

70. Pai BC, Ramani G, Pillai RM, Satyanarayana KG (1995) Review: role of magnesium in cast aluminum alloy matrix composites. J Mater Sci 30:1903–1911

71. Shoutens JE (1992) Some theoretical considerations of the surface tension of liquid metals for metal matrix composites. J Mater Sci 24:2681–2686

72. Contreras A (2007) Wetting of TiC by Al–Cu alloys and interfacial characterization. J Colloid Interface Sci 311:159–170

73. Lloyd DJ (1991) Aspects of fracture in particulate reinforced metal matrix composites. Acta Metall Mater 39:59–71

74. Ravi-Kumar NV, Dwarakadasa ES (2000) Effect of matrix strength on the mechanical properties of Al-Zn-Mg/SiC$_p$ composites. Compos Part A 31:1139–1145

75. Fine ME, Conley JG (1990) On the free energy of formation of TiC and Al$_4$C$_3$. Metall Trans 21A:2609–2610

76. Yokokawa H, Sakai N, Kawada T, Dakiya M (1991) Chemical potential diagram of Al-Ti-C system: Al$_4$C$_3$ formation on TiC formed in Al-Ti liquids containing carbon. Metall Trans 22A:3075–3076

77. Kennedy AR, Weston DP, Jones MI (2001) Reaction in Al-TiC metal matrix composites. Mater Sci Eng A 316:32–38

78. Frage N, Frumin N, Levin L, Polak M, Dariel MP (1998) High-temperature phase equilibria in the Al-rich corner of the Al-Ti-C system. Metall Mater Trans A 29:1341–1345

79. Samuel AM, Gauthier J, Samuel FH (1996) Microstructural aspects of the dissolution and melting of Al$_2$Cu phase in Al-Si alloys during solution heat treatment. Metall Mater Trans A 27:1785–1798

80. Aguilar EA, Leon CA, Contreras A, Lopez VH, Drew RAL, Bedolla E (2002) Wettability and phase formation in TiC/Al-alloys assemblies. Compos Part A 33:1425–1428

81. López VH, Leon CA, Kennedy A et al (2003) Spreading mechanism of molten Al-alloys on TiC substrates. Mater Sci Forum 416–418(3):395–400

82. Leon CA, Lopez VH, Bedolla E, Drew RAL (2002) Wettability of TiC by commercial aluminum alloys. J Mater Sci 37:3509–3514

83. Albiter A, Contreras A, Salazar M, Gonzalez JG (2006) Corrosion behaviour of aluminium metal matrix composites reinforced with TiC processed by pressureless melt infiltration. J Appl Electrochem 36:303–308

84. Duran-Olvera JM, Orozco-Cruz R, Galván-Martínez R, León CA, Contreras A (2017) Characterization of TiC/Ni composite immersed in synthetic seawater. MRS Adv 2(50):2865–2873

85. Alvarez-Lemus N, Leon CA, Contreras A, Orozco-Cruz R, Galvan-Martinez R (2015) Chapter 15: Electrochemical characterization of the aluminum–copper composite material reinforced with titanium carbide immersed in seawater. In: Perez R, Contreras A, Esparza R (eds) Materials characterization. Springer, Cham, pp 147–156

86. Lugo-Quintal J, Díaz-Ballote L, Veleva L, Contreras A (2009) Effect of Li on the corrosion behavior of Al-Cu/SiC$_p$ composites. Adv Mater Res 68:133–144

87. Santamaria D (2001) Efecto del tratamiento térmico de solución y precipitación a un material compuesto de matriz metálica TiC/Al-6061. Dissertation of Master Thesis, Instituto de Investigación en Metalurgia y Materiales, UMSNH, Morelia, México

88. Harris GL (1995) Properties of silicon carbide. Materials Science Research Center of Excellence. Howard University, Washington DC, p 304
89. Snead LL (2004) Limits on irradiation-induced thermal conductivity and electrical resistivity in silicon carbide materials. J Nucl Mater 329–333:524–529
90. Wang H, Zhang R, Hu X et al (2008) Characterization of a powder metallurgy SiC/Cu–Al composite. J Mater Process Technol 197:43–48
91. Kocjak M et al (1993) Fundamentals of metal matrix composites. Blutterworth-Heinemann, Waltham, pp 3–42
92. Chu K, Jia C, Tian W et al (2010) Thermal conductivity of spark plasma sintering consolidated SiC_p/Al composites containing pores: numerical study and experimental validation. Compos Part A 41:161–167
93. Chen Q, Yang W, Dong R et al (2014) Interfacial microstructure and its effect on thermal conductivity of SiCp/Cu composites. Mater Des 63:109–114
94. Hasselman DPH, Johnson LF (1987) Effective thermal conductivity of composites with interfacial thermal barrier resistance. J Compos Mater 21:508–515
95. Beffort O, Long S, Cayron C et al (2007) Alloying effects on microstructure and mechanical properties of high volume fraction SiC-particle reinforced Al-MMCs made by squeeze casting infiltration. Compos Sci Technol 67:737–745
96. Jae-Chu L, Ji-Young B, Sung-Bae P et al (1998) Prediction of Si contents to suppress the formation of Al_4C_3 in the SiC_p/Al composite. Acta Mater 46(5):1771–1780
97. Ren S, He X, Qu X et al (2007) Effect of Mg and Si in the aluminum on the thermo-mechanical properties of pressureless infiltrated SiCp/Al composites. Compos Sci Technol 67 (10):2103–2113
98. Rajan T, Pillai R, Pai B (1998) Reinforcement coatings and interfaces in aluminium metal matrix composites. J Mater Sci 3:3491–3503
99. Kim Y, Lee J (2006) Processing and interfacial bonding strength of 2014 Al matrix composites reinforced with oxidized SiC particles. Mater Sci Eng A 420:8–12
100. Xue C, Yu J (2014) Enhanced thermal transfer and bending strength of SiC/Al composite with controlled interfacial reaction. Mater Des 53:74–78
101. Zalapa O (2016) Síntesis y evaluación de propiedades termofísicas de compuestos de matriz de Mg-AZ91E reforzados con partículas de SiC. Dissertation of Master Thesis, Instituto de Investigación en Metalurgia y Materiales, UMSNH, México
102. Ureña A et al (2004) Oxidation treatments for SiC particles used as reinforcement in aluminium matrix composites. Compos Sci Technol 64(12):1843–1854
103. Kerner EH (1956) The elastic and thermo-elastic properties of composite media. Proc Phys Soc 69:808
104. Basavarajappa S, Chandramohan G, Mahadevan A (2007) Influence of speed on the dry sliding wear behavior and subsurface deformation on hybrid metal matrix composite. Wear 262:1007–1012
105. Prakash K, Balasundar P, Nagaraja S et al (2016) Mechanical and wear behaviour of Mg-SiC-Gr hybrid composites. J Magnes Alloys 4:197–206
106. Sozhamannan G, Balasivanandha S, Venkatagalapathy V (2012) Effect of processing parameters on metal matrix composites: stir casting process. J Surf Eng Mater Adv Technol 2:11–15
107. Arreola C (2017) Evaluación de propiedades mecánicas y comportamiento al desgaste de compuestos AZ91E/AlN fabricados por fundición con agitación. Dissertation of Master Thesis, Instituto de Investigación en Metalurgia y Materiales, UMSNH, México
108. Grabowski G, Pedzich Z (2007) Residual stresses in particulate composites with alumina and zirconia matrices. J Eur Ceram Soc 27:1287–1292
109. Gutknecht D, Chevalier J, Garnier V et al (2007) Key role of processing zirconia composites for orthopedic application. J Eur Ceram Soc 27:1547–1552
110. Nakao E, Ono M, Lee SK et al (2005) Critical crack-healing condition for SiC whisker reinforced alumina under stress. J Eur Ceram Soc 25:3649–3655

111. Yang JF, Ohji T, Sekino T et al (2001) Phase transformation, microstructure and mechanical properties of Si3N4/SiC composite. J Eur Ceram Soc 21(12):2185–2192
112. Sekino T, Nakajima T, Ueda S et al (1997) Reduction and sintering of a nickel-dispersed-alumina composite and its properties. J Am Ceram Soc 80:1139–1148
113. Wada S, Suganuma M, Kitagawa Y et al (1999) Comparison between pulse electric current sintering and hot pressing of silicon nitride ceramics. J Ceram Soc Jpn 107(10):887–890
114. Xie G, Ohashi O, Sato T et al (2004) Effect of Mg on the sintering of Al-Mg alloy powders by pulse electric current sintering process. Mater Trans 45(3):904–909
115. Dang KQ, Nanko M, Kawahara M et al (2009) Densification of alumina powder by using PECS process with different pulse electric current wave forms. Mater Sci Forum 620–622:101–104
116. Suk MJ, Choi SI, Kim JS et al (2003) Fabrication of a porous material with a porosity gradient by a pulsed electric current sintering process. Met Mater Intern 9(6):599–603
117. Xie G, Ohashi O, Yamaguchi N (2004) Reduction of surface oxide films in Al–Mg alloy powders by pulse electric current sintering. J Mater Res 19(3):815–819
118. Matsubara T, Shibutani T, Uenishi K et al (2000) Fabrication of a thick surface layer of Al₃Ti on Ti substrate by reactive-pulsed electric current sintering. Intermetallics 8:815–822
119. Salas-Villaseñor AL, Lemus-Ruiz J, Nanko M et al (2009) Crack disappearance by high-temperature oxidation of alumina toughened by Ni nano-particles. Adv Mater Res 68:34–43
120. Salas-Villaseñor AL (2008) Auto-eliminación de grietas por oxidación a elevada temperatura de alúmina reforzada con níquel. Dissertation of Master Thesis, Instituto de Investigación en Metalurgia y Materiales, UMSNH, Morelia, México
121. Niihara K, Kim BS, Nakayama T et al (2004) Fabrication of complex-shaped alumina/nickel nanocomposites by gel casting process. J Eur Ceram Soc 24:3419–3425
122. Lu J, Gao L, Sun J et al (2000) Effect of nickel content on the sintering behavior, mechanical and dielectric properties of Al₂O₃/Ni composites from coated powders. Mater Sci Eng A 293:223–228
123. Lieberthal M, Kaplan WD (2001) Processing and properties of Al2O3 nanocomposites reinforced with sub-micron Ni and NiAl₂O₄. Mater Sci Eng A 302:83–91
124. Tuan WH (2005) Design of multiphase materials. Key Eng Mater 280–283:963–966
125. JIS R-1607 Japanese Industrial Standard (1990) Testing methods for fracture toughness of high performance ceramics. Japanese Standards Association, Tokyo
126. Miyoshi T, Sagawa N, Sassa T (1985) Study on fracture toughness evaluation for structural ceramics. Trans Jpn Soc Mech Eng 51A(471):2487–2489
127. Casellas D, Nagl MM, Llanes L et al (2003) Fracture toughness of alumina and ZTA ceramics: microstructural coarsening effects. J Mater Process Technol 143–144:148–152
128. Reyes A, Bedolla E, Perez R, Contreras A (2016) Effect of heat treatment on the mechanical and microstructural characterization of Mg-AZ91E/TiC composites. Compos Interfaces:1–17
129. Reyes A (2012) Caracterización interfacial del compuesto MgAZ91E/TiC con y sin tratamiento térmico. Dissertation of Master Thesis, Instituto de Investigación en Metalurgia y Materiales, UMSNH, Morelia, México
130. Munitz A, Jo I, Nuechterlein J, Garrett W, Moore JJ, Kaufman MJ (2012) Microstructural characterization of cast Mg-TiC MMC's. Int J Mater Sci 2:15–19
131. Contreras A, Albiter A, Pérez R (2004) Microstructural properties of the Al-Mg/TiC composites obtained by infiltration techniques. J Phys Condens Matter 16(22):S2241–S2249
132. Halpin JC, Kardos JL (1976) The Halpin-Tsai equations: a review. Polym Eng Sci 16 (5):344–352
133. Xiuqing Z, Lihua L, Naiheng M, Haowei W (2006) Effect of aging hardening on in situ synthesis magnesium matrix composites. Mater Chem Phys 96(1):9–15

Chapter 5
Joining of Composites

5.1 Introduction

Metal matrix composites (MMC) represent a class of material with special characteristics that make them functional for a wide range of practical applications in many areas of daily life such as in sport [1], recreation [2], and electronics industry [3], as well as structural applications, as automotive industry [4] where high temperature and mechanical properties are required, such as wear [5]. Metal composite material applications make use of composite materials. These materials are produced in situ from the conventional production and processing of metals, as well as structures, which result from welding of steel. Materials like cast iron with graphite or steel with high carbide content, as well as tungsten carbides, consisting of carbides and metallic binders, also belong to this group of composite materials [6]. Growth in the development of MMC opens up unlimited possibilities for modern material science into the most important applications. The possibility of combining various material systems (metal/ceramic/nonmetal) gives the opportunity for unlimited variation. The properties of these new materials are basically determined by the properties of their single components.

The reinforcement of light metals opens up the possibility of application of these materials in areas where weight reduction has first priority, but must improve the component properties. The development objectives for MMC are the following: increase in resistance at higher temperatures compared to that of conventional alloys, improvement of thermal shock and corrosion resistance, and increase in Young's modulus [7]. For other applications, different development objectives are given, which differ from those mentioned before. For example, in Aerospatiale applications, while the desire for high-precision, dimensionally stable spacecraft structures has driven the development of MMCs, applications thus far have been limited by difficult fabrication processes [8]. The extensive use of metallic alloys in engineering reflects not only their strength and toughness but also the relative ease and low cost

© Springer Nature Switzerland AG 2018
A. Contreras Cuevas et al., *Metal Matrix Composites*,
https://doi.org/10.1007/978-3-319-91854-9_5

Table 5.1 Qualitative rating for joining adaptability, applications, and selection

Joining method	Joint applications			Adaptability for MMC
	High strength	High temperature	Complex shapes	
Friction stir welding	*	*	−	*
Ultrasonic welding	*	*	+	*
Diffusion bonding	+	*	−	+
Transient liquid phase	+	*	*	+
Rapid infrared joining	+	*	+	+
Laser beam welding	*	*	+	−
Electron beam welding	*	*	+	−
Gas metal arc welding	*	*	*	−
Gas tungsten arc welding	*	*	*	−
Resistance spot welding	*	*	+	−
Capacitor discharge welding	*	*	+	+
Brazing	+	+	*	+
Soldering	+	−	*	*
Adhesive bonding	+	−	+	*
Mechanical fastening	*	*		*

Joint performance rating: * high, + medium, − low

of fabrication of engineering components by a wide range of manufacturing processes.

The development of MMCs has reflected the need to achieve property combinations beyond those attainable in monolithic metals alone. For the most practical applications of the MMC, use of the MMC in combination with a metal or a light alloy is necessary; this makes joining process a technological key to expand the use of MMC [3, 9]. Several methods to joining dissimilar materials have been developed; however, sometimes it is not possible make the adaptations necessary to joints MMC. Table 5.1 presents the potential applications of different joining methods and the possible adaptation or modifications to be used during joining of MMC [4, 10].

It is important to understand that MMC joining is not a modern technology, and many important joining technical details are still deficient. As a result, the precise knowledge of the adaptability for a specific joining method is a specific material and process-dependent factor which must be determined experimentally. On the other hand, as a general observation, the use of solid-state and other low-temperature processes is often more adaptable for joining of MMCs than the use of high-temperature fusion processes. However, selecting a joining method can be done using a set of criteria for joint applications, such as joint strength, thermal and electrical conductivity, etc., in conjunction with its adaptability for joining MMCs. Further details of some process and classification are provided in subsequent sections.

5.2 Classification of MMC Joining Processes

While the extensive uses of metallic alloys in engineering reproduce not only their strength and toughness but also the relative and low cost of fabrication of engineering components by a wide range of manufacturing processes, the potential applications of different joining methods depend in the possible adaptation or modifications to be used during joining of MMC. Successful applications of advanced materials in many devices and structures require some type of dissimilar joining. Using joining processes as part of a manufacturing route can offer considerable technical and economic advantages to the designer, provided careful and informed decisions are made about the processes to be applied, materials to be selected, and joint configurations and the process parameters to be employed.

There are many reasons for wishing to join particular ceramic and metal components, but the motives can usually be related to design, manufacturing, or economic factors. The development of viable joining techniques will facilitate the assimilation of advanced materials into complex multimaterial structures. Several methods have been developed over the years to produce hardy and reliable joints between MMC. The choice of appropriate joining method depends on the materials to be joined, the joint design, and the anticipated service conditions. Several methods to joining dissimilar materials have been developed; in Fig. 5.1, a classification of different bonding processes in function of their nature could be found [4, 11].

There are two general categories of joining processes: i) diffusion processes which is separated in solid-state (direct) and liquid-state (indirect) processes, and ii) fusion processes, in which components are bonded by fusion. Some authors are mention; direct joining, in which components are bonded without an interlayer of material, and indirect joining, in which an intermadiate layer of material, such as an adhesive, cement or braze is used to bond the components [12, 13].

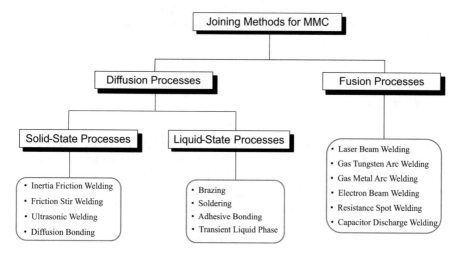

Fig. 5.1 Classification of selected joining methods for MMCs

Successful solid-state or direct joining relies upon the achievement of adequate interfacial contact and subsequent diffusion or plastic flow to eliminate interfacial porosity. When both materials undergo limited plastic flow, e.g., MMC-ceramic bonding, or when deformation of the workpiece must be avoided, special care must be taken to assure smooth mating surfaces [14]. The advantages of solid-state or direct bonding include a simple fabrication procedure, a one-step process, and potentially high joint strength. However, there are also several limitations and disadvantages: it is of high cost; only relatively planar specimens can be joined; a vacuum/inert atmosphere is required; and pressure must be applied. The need to apply pressure during diffusion bonding imposes restrictions on the joint geometry; most joints are of the face seal type and are not well suited for accommodating thermal expansion mismatch [15]. Whatever process is used, the formation of successful joints depends on the achievement of intimate contact between the work pieces, the conversion of these contacting surfaces into an atomically bonded interface, and the ability of this interface to accommodate thermal expansion mismatch stresses generated during cooling process after fabrication or temperature changes during operational conditions.

In liquid-state joining, the use of a liquid, a glass, or a solid foil that flows readily under low applied stress to join materials can have advantages. Flow of a wetting liquid can fill irregularities in the surface and therefore imposes less stringent demands on surface preparation and the degree or extent of surface mating required. Indirect joining is the most common method of achieving high integrity joints using a wide range of intermediate bonding materials. The major categories of joining using an intermediate layer include joining with adhesives, cements, glasses or glass-ceramics, and brazes. Joining with organic-based materials is simple and widely used. A broad range of materials can be joined by this method. Precise processing conditions have been specified for the preparation of aluminum and aluminum alloy surfaces for adhesive bonding, a consequence of the importance of this joining approach in the aerospace industry. The indirect bonding of ceramics or MMC includes those techniques in which a liquid medium is responsible for bonding.

5.2.1 *Friction Welding*

Friction-based processes have advantages that include low distortion and thermal disruption and the ability to join dissimilar materials. Friction welding is a way of joining both similar and dissimilar combinations of materials. It is process that is well established and widely accepted throughout industry. The weld is made by rotating one component against a fixed component under pressure. This action efficiently cleans the weld faces and generates sufficient frictional heat for at least one of the materials to become plastic at the joint interface. This joining process produces a merging between the metals and nonmetals by using the heat developed between the two surfaces. This condition occurs in combination with the mechanically induced rubbing motion and an applied load. At this point, rotation is stopped rapidly and the components are forged together. The result is a highly effective bond over the whole joint area; in most cases the joint strength is equal in strength to that of the parent metals. Under normal conditions, the merging materials melt to form this bond; this is also accomplished without the use of filler metals, fluxes, or shielding gases. Rotation speed, friction time, friction and forge pressures, and length loss must all be carefully controlled. These are the basic process parameters, and vary for different components and materials. Generally, at least one component must be circular at the joint face [16, 17].

The many advantages of friction welding include substantial labor savings, high production rates, joint strength equal to or greater than parent metal, self-cleaning action, reduction or elimination of surface prep, very reliable integrity of welded joints, highly precise and repeatable process, possible joint dissimilar metal combinations, environmentally friendly, and no filler metals, fluxes, or protective gases needed [18, 19]. Rotary friction welding is an extremely reliable and cost-effective method of joining metal parts. It is a well-established technology with applications in industries such as automotive [20] and aerospace [21], to name a few. The process of friction welding, in practice, is accomplished in several phases represented in Fig. 5.2.

1. *Before welding*: First parts are mounted in the friction welder, and rotating part is spun up to a speed of 1000 RPM.
2. *Phase 1* (first friction): Parts are rubbing together, at low force, to accomplish a cleanup of the two surfaces to be welded; the force applied during first friction is ~30% of the second friction; *Phase 2* (second friction): The increased pressure brought about during second friction causes the metal to become *plastic* and flows

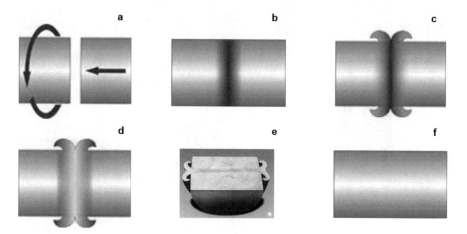

Fig. 5.2 Stages during friction welding process: (**a**) before welding, (**b**) Phase 1 (first friction), (**c**) Phase 2 (second frictions), (**d**) Phase 3 (forge), (**e**) finished weld, (**f**) flash removal [18, 19]

outward from center to form the characteristic *flash;* once the designed *flash* is accomplished, the rotation is rapidly stopped. The process then moves to the forge phase.

3. *Phase 3* (forge): The forge is caused by the application of the highest of the three process pressures. The forge phase takes place while the components are at a complete stop. The pressure is maintained until the weld joint is sufficiently cooled, this step promotes refinement of the microstructure of the weld. Friction welding produces forged quality joints, with 100% butt joint weld through the contact area. This solid section shows the narrow heat-affected weld zone and resulting displaced material.

4. *Flash removal*: The *flash* is removed (if desired) by conventional machining practices. Removal of the weld *flash* is optional.

Figure 5.3 shows a friction welding machine and welding process. A typical friction welding machine consists of friction welding machine head, machine base, component clamping arrangement and backstop, hydraulic power supply, electrical/electronic control, automatic machine lubrication, and machine monitoring device

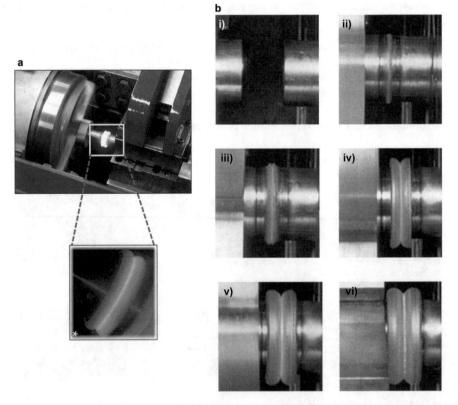

Fig. 5.3 (**a**) Friction welding machine and (**b**) friction welding process: (i) before welding, (ii) early phase, (iii) middle phase, (iv) late middle phase, (v) late phase, (vi) finished weld [22]

(optional) [22]. The machine designer has the choice to make as to whether the friction welding machine head moves to a fixed component clamping arrangement or the reverse. In the friction welding process, one component is commonly held in a self-centering clamp. The most commonly used friction welding machine joins two components together, although three or more components can be simultaneously joined on a specially designed machine.

The machine spindle can be driven directly from an AC or DC motor source and then allowed to stop under its natural deceleration characteristics, and the retarding torque developed from the components being formed. The variable parameters of the friction welding process are as follows: (a) speed (only when DC drive is used), (b) pressure, and (c) loss of length (or time). For a particular application, heating time is determined during the setup or from previous experience; the friction pressure depends on the materials being joined and the surface joint geometry; the rotational speed is related to the diameter of the welding material; an appropriate rotation speed must be used to minimize any harmful effects and produce good quality of joints [23].

5.2.2 Welding Process

Welding involves the fusion of the joint surface by controlled melting through heat being specifically directed toward the joint. Heating sources commonly used include plasma arcs, electron beams, and electrical current through the components and across the joints; however, the electric arc is the power source employed in most welding processes as heat source to accomplish fusion. Welding processes can be classified into two main categories: (1) liquid-phase welding, e.g., all fusion welding processes such as conventional arc welding, laser welding, and electron beam welding [24, 25], and (2) solid-state welding, e.g., forge welding, friction stir welding, explosive welding, and solid-sate diffusion bonding [26, 27]. In the first case, bonds are established by the formation and solidification of a liquid phase at the interface, while, in the latter case, the applied pressure has a key role in bringing together the surfaces to be joined within inter atomic distances.

The main characteristic features of welding include the following: heating cycle must be rapid, and it usually affects the microstructure and hence the properties of the components over a macroscopic region around the joint called the heat-affected zone (HAZ); joint geometries are limited by the requirement that all joint surfaces are accessible to the concentrated heat source; welded joints may approach the physical integrity of the component, but are often inferior in their mechanical properties; it is due to stress concentrations produced by the high thermal gradient developed during joining; the HAZ is often influential in determining the properties of welded joints; and welding tends to distort the components in the region of the HAZ due to the use of a concentrated heat source. Filler metal may be used to supplement the fusion process for components of similar composition, for example, when the joint gap is wide and of variable width as is shown in Fig. 5.4.

Fig. 5.4 Joint gap for welding engineering materials

Various types of welding processes according to their source of heat application include gas (oxy-acetylene, air-acetylene), resistance (gas metal arc, plasma, gas tungsten, carbon arc), and electron beam and laser beam welding [26]. The use of Al-based composites is still limited by their poor weldability when using conventional fusion welding processes, which produce a weld pool that has poor fluidity due to the presence of the ceramic phase and solidifies with a large volume of porosity in both the weld and HAZ because hydrogen evolves from the aluminum matrix, particularly, in MMCs. High temperatures, typically above the liquidus of the aluminum alloys, result in a severe degradation of the mechanical properties caused by the formation of brittle and hydroscopic aluminum-carbon compounds, mainly aluminum carbide, as is the case of Al/SiC-type composite irrespective of the nature of the welding process employed [28, 29].

The direct establishment of the electric arc during the welding process of metallic materials gives rise to full exposure of the materials to the elevated temperatures developed by the process. Although all metallic materials can be joined by the MIG welding process, the penetration is not enough to weld in one pass a thickness of 12.5 mm in aluminum, and therefore a multi-pass procedure is required, giving rise to wide and nonuniform welds as a result of the significant volume of metal that has to be fed to fill the groove of the joint. MIG-IEA welding process proposes to make good use of the thermal energy generated by the electric arc to succeed fusion and welding in one pass that has been studied for thick aluminum plates [30].

Garcia et al. [31] have reported that welding of Al-based composites reinforced with a high ceramic content of TiC (50%) particles was possible by the MIG process using both direct (DEA) and indirect (IEA) electric arc techniques. They found that a complete penetration was achieved in one welding pass by preheating the parent composite above 50 °C, but a double welding pass was needed to weld the composite by DEA at room temperature. While areas free of dispersed particles and uniform welds are obtained by IEA, wide weld beads in the arc applied region with some incorporation of the reinforcing particles are produced by DEA, as can be observed in Fig. 5.5. Evaluation of the mechanical efficiency in the welds indicates that, independently of the welding technique and preheating condition employed, TiC-reinforced aluminum MMCs exhibited a good degree of weldability by the MIG process.

Although the efficiencies exhibited by IEA are lower than by DEA, the former looks attractive due to the weld profiles that can be obtained with a reduced HAZ, as can be observed in Fig. 5.6. Thus, the use of the MIG welding process with IEA can be considered an alternative for solving some problems encountered in welding of MMCs. Since the defects of the weld deposits by IEA are located outside of the weld bead in the upper part, sound welds may be obtained by this modified MIG welding

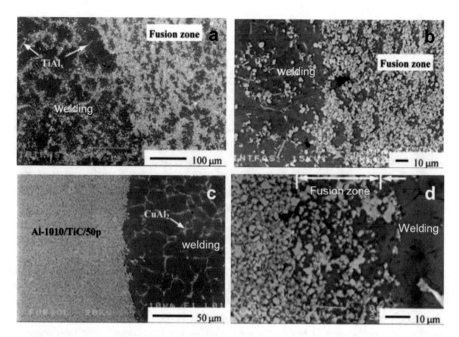

Fig. 5.5 Microstructure of the welds preheated at 150 °C and produced by (**a**) and (**b**) DEA and (**c** and **d**) IEA [31]

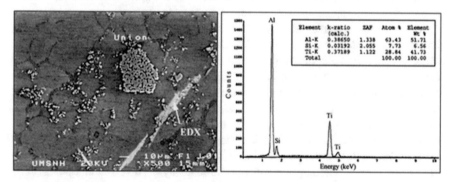

Fig. 5.6 Detail of the HAZ in a weld obtained by DEA at 150 °C and EDX analysis of the acicular phase [31]

process. The improvement of the efficiency of the MIG welding process up to 94.8% using IEA may be ascribed to the better use of heat generated by the electric arc, which is established in a hidden form, and the contact with the environment is minimum reducing, thus heat losses. These results are reported by Garcia et al. [32] during welding of aluminum by the MIG process with indirect electric arc (MIG-IEA).

5.2.3 Diffusion Bonding

Diffusion bonding, as a subdivision of both solid-state and liquid-phase bondings, is a joining process in which the principal mechanism is interdiffusion of atoms across the interface. Diffusion bonding of most metals is conducted in a vacuum or in an inert atmosphere (normally dry nitrogen, argon, or helium) in order to reduce detrimental oxidation of the faying surfaces. Bonding of a few metals which have oxide films that are thermodynamically unstable at the bonding temperature (e.g., silver) may be achieved in air.

There are two possible mechanisms of the solid-phase bonding without the intermediary of a liquid filler metal: when the solid-phase MMC and metal are brought into contact, the metal may undergo plastic deformation, enter the surface irregularities of the MMC, and adhere and bond to the MMC; in addition diffusion can occur between the metallic matrixes of the MMC with the monolithic bonding metal. In another possible mechanism, the metal may diffuse through the interface, react with the ceramic and form a continuous layer [15, 33]. Diffusion bonding has attracted interest as a means of bonding and successes have been achieved by controlling the microchemistry and microstructure of the interfaces formed. The first requirement for solid-state diffusion bonding is the creation of intimate contact between the surfaces to be bonded in order that the atomic species come into intimate contact. In addition to a good contact, there should be enough diffusion between the materials in a reasonable time. Pressure can be applied uniaxially (hot-press) or isostatically (hot isostatic press) on a diffusion couple. Figure 5.7 shows an illustration of events during solid-state diffusion bonding [11, 14].

There are several advantages of solid-state joining. When a MMC/MMC has to be bonded, the contact areas between the couple ensemble must be metal/metal and ceramic-metal. It is a common practice to introduce a metal interlayer between the components. The interlayer should be ductile so that it can deform readily to achieve intimate contact with both mating surfaces at various pressures and temperatures and of course that it should adhere strongly to both the metal and ceramic components. The technological advantages of diffusion bonding are low deformation which enables parts to be joined with little distortion, the ability to join large areas, the applicability of diffusion-bonded joints at high service temperatures, and the possibilities for joining materials in a nonconventional way [34]. The technique has been applied, so far, mainly for the joining of MMC, but its utility has also been demonstrated for joining other new engineering ceramics like WC.

Some limitations are associated to diffusion bonding, mainly when the process is carried out at solid-state bonding, where (1) great care is required in the surface preparation stage. Excessive oxidation or contamination of the faying surfaces would decrease the joint strength drastically. Diffusion bonding of materials with stable oxide layers is very difficult. Production of thoroughly flat surfaces and also precise fitting up of the mating parts take a longer time than with conventional welding processes. (2) The initial investment is fairly high, and production of large components is limited by the size of the bonding equipment used, and (3) the suitability of this process for mass production is questionable, particularly because of the long bonding times involved [35].

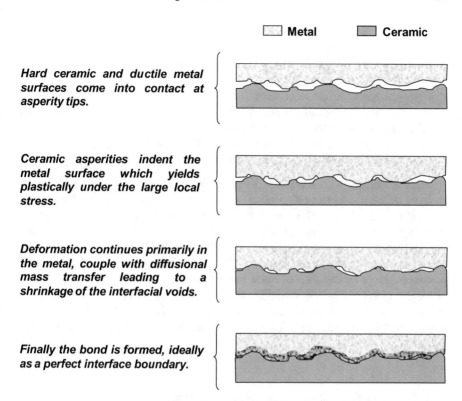

Hard ceramic and ductile metal surfaces come into contact at asperity tips.

Ceramic asperities indent the metal surface which yields plastically under the large local stress.

Deformation continues primarily in the metal, couple with diffusional mass transfer leading to a shrinkage of the interfacial voids.

Finally the bond is formed, ideally as a perfect interface boundary.

Fig. 5.7   Sequence of events during solid-state diffusion bonding [11]

5.2.4   Solid-State Diffusion Bonding

Solid-state diffusion bonding is a process by which two nominally flat interfaces can be joined at an elevated temperature (about 50–90% of the absolute melting point of the parent material) using an applied pressure for a time ranging from a few minutes to a few hours. The International Institute of Welding (IIW) has adopted a modified definition of solid-state diffusion bonding: "Diffusion bonding of materials in the solid state is a process for making a monolithic joint through the formation of bonds at atomic level, as a result of closure of the mating surfaces due to the local plastic deformation at elevated temperature which aids interdiffusion at the surface layers of the materials being joined."

Some advantages of solid-state diffusion bonding are [36, 37]:

- The process has the ability to produce high-quality joints so that neither metal-lurgical discontinuities nor porosity exists across the interface.
- With properly controlled process variables, the joint would have strength and ductility equivalent to those of the parent material.

- Joining of dissimilar materials with different thermophysical characteristics, which is not possible by other processes, may be achieved by diffusion bonding. Metals, alloys, ceramics, and MMC products have been joined by diffusion bonding [38].
- High-precision components with intricate shapes or cross sections can be manufactured without subsequent machining. This means that good dimensional tolerances for the products can be attained.
- Diffusion bonding is free from ultraviolet radiation and gas emission so there is no direct detrimental effect on the environment.

Some problems with solid-state diffusion bonding are [11, 39]: to bring the surfaces of the two pieces being joined sufficiently close so that interdiffusion can result in bond formation. There are two major obstacles that need to be overcome in order to achieve satisfactory diffusion bonds: (1) polished surfaces come into contact only at their asperities, and (2) hence, the ratio of contacting area to faying area is very low. Solid-state diffusion bonding, also called direct bonding, is a solid-phase process achieved by diffusion of atoms that take place at the interface. In the absence of a liquid phase, the joint has to be made at high temperatures, prolonged times, and using pressure to produce intimate contact and to provide sufficient thermal energy to permit diffusion and chemical reaction between the matrix, ceramic, and metallic surfaces.

5.2.4.1   Effect of Bonding Temperature and Time

Temperature is the most important parameter in the bonding process due to the fact that (1) in thermally activated processes, a small change in temperature will result in the greatest change in process kinetics, diffusion, and creep, compared with other parameters, and (2) virtually all mechanisms in diffusion bonding are sensitive to temperature, plastic deformation, diffusion, and creep. Temperature increases interaction across an interface by increasing the mobility of atoms and also the mobility of dislocations in the metal during bonding. Since the mobility of dislocations increases with temperature and the flow stress correspondingly decreases, the pressure required for bonding decreases with increasing temperature. In general, the temperature required to obtain sufficient joint strength is typically within the range *0.6 and 0.95 Tm*, where *Tm* is the absolute melting point of the base material [11].

In order to clarify the effect of bonding temperature and time during solid-state bonding, two cases will be analyses. At the beginning, Bedolla et al. [40] makes complete description of the synthesis and characterization of metal matrix composites (MMCs) of magnesium alloy AZ91E reinforced with AlN (49 vol.%). Molten Mg alloy (AZ91E) was pressureless infiltrated at 900 °C into AlN preforms sintered at 1450 °C with porosities around 51 vol.%. The composites obtained were microstructurally characterized, and SEM observations showed a homogeneous microstructure of matrix and reinforcement. AlN, Mg, and $Mg_{17}Al_{12}$ phases were detected through X-ray diffractions. The presence of $Mg_{17}Al_{12}$ is a typical second

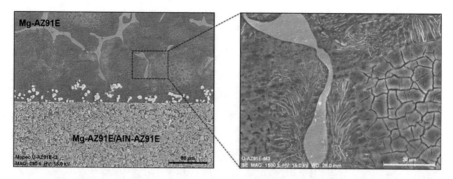

Fig. 5.8 SEM image of the cross-sectional bonding of Mg-AZ91E/AlN-AZ91E combinations joined at 600 °C for 5 h [41]

phase of the Mg alloy (AZ91E). In general, MMCs utilize a variety of nonmetallic reinforcements with a typical volume fraction ranging from 5% to 60%. For this reason, there are a number of potential joining issues that are peculiar to MMCs.

1. *In the first case*, a MMC/AZ91E combination was produced by solid-state diffusion bonding of AlN-AZ91E composite joint to monolithic AZ91E alloy at 600 °C for 5 h under argon atmosphere. It is possible to observe a cross section of the interface produced in this joint by scanning electron microscopy (SEM) which is showed in Fig. 5.8 [41].

   Fig. 5.8 shows the following characteristics: a homogeneous bonding of Mg-AZ91E alloy to AlN-AZ91E MMC material at 600 °C for 5 h by solid diffusion bonding, joint combination formed through the diffusion of Mg alloy components with the Mg alloy in the matrix of the MMC, a continuous and homogeneous diffusion zone free of porosity and thermal cracks, the formation of spherical precipitates inside the layer near the bonding line with the MMC, and the amount of these precipitates which increase with time. In general, the joining process temperature and time must be carefully controlled such that the contact between molten metal matrix and the reinforcements during joining will not lead to dissolution of the reinforcement material, interdiffusion, and the formation of undesirable metallurgical phases. The chemical stability of the metal matrix-reinforcement for a joining method depends of the specific material and process; consequently, final joining parameters for a specific process must be experimentally determined. On the other hand, the temperature and time effect can be observed in Fig. 5.8, and embrittlement of the Mg-AZ91E alloy can be present by the heat treatment produced during joining with the migration of Mn commonly present close to the $Mg_{17}Al_{12}$ precipitate of the Mg-AZ91E alloy, near the bonding line of the MMC combinations. Since solid-state joining of Mg-AZ91E alloy to metal matrix composite (AlN-AZ91E) occurred at relatively high temperature (close to melting point of the Mg alloy) during the process, this promoted interfacial interactions and bonding between the materials. A qualitative overview of the different components across the Mg-AZ91E/AlN-AZ91E interface was studied using atomic distributions for a

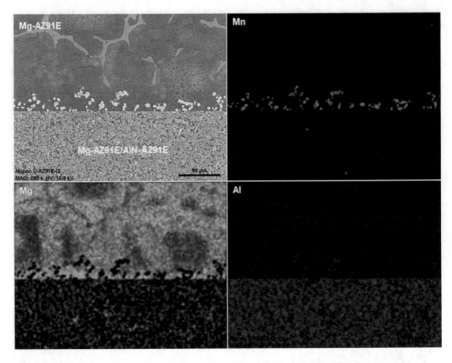

Fig. 5.9 Mapping analyses through the cross section of Mg-AZ91E/AlN-AZ91E combinations joined at 600 °C for 5 h [41]

sample joined at 600 °C for 5 h. The results are illustrated in Fig. 5.9, where the interface is aligned with the Mg-AZ91E on the top and the composite AlN-AZ91E on the underside.

The main elements analyzed were Mg, Al, and Mn. The different contrasts from dark to colored correspond to the increase in the concentration of the specific element. In the Mg map, the different contrasts correspond to the decrease in concentration of Mg corresponding to the $Mg_{17}Al_{12}$ second phase inside the AZ91E alloy and matrix inside the MMC. On the other hand, similar contrast is observed for the Al map, and diffusion of Mn can be observed on the Mn map. It must be observed that the intensity of color of the spherical precipitates inside the layer demarks clearly the diffusion of Mn and Al components; however, Mg was not observed inside these spherical precipitates. The EPMA-EDS carried out in these samples shows that Mn and Al are the main elements in the spherical precipitate phase, which increases in amount when the bonding time increased. An overview of the different components across the Mg-AZ91E/AlN-AZ91E interface was obtained using element line scan analyses of the joints bonded at 600 °C for 5 h; the results are illustrated in Fig. 5.10.

The elements analyzed were Al, Mg, Mn, Si, O, and N. The analysis was performed starting on the MMC side of the sample through the bonding interface

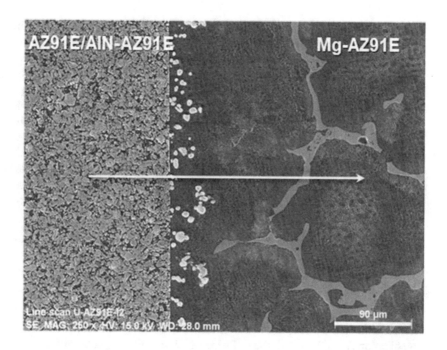

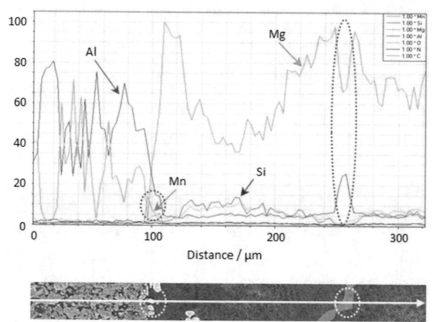

Fig. 5.10 Electron probe line microanalysis through the cross section of Mg-AZ91E/AlN-AZ91E combinations joined at 600 °C for 5 h [41]

passing through the different reaction phases up to the pure Mg-AZ91E alloy. In the case of the Mn composition profile, it was possible to observe its diffusion inside the interface and obtain its concentration in the spherical Mn-rich precipitates. No diffusion of Si into the MMC was observed, resulting in the drop of the Si concentration profile. A concentration decrease can be observed for the Mg and increase of Al profiles inside the $Mg_{17}Al_{12}$ precipitates, observing the high concentration profiles at the Mg-AZ91E alloy and MMC for Mg and Al, respectively. Diffusion is the dominating reaction mechanism in diffusion joining; consequently the high affinity of the Mg alloy of the MMC with the Mg-AZ91E alloy resulted in immediate diffusion with some modifications depending on the joining parameters, such as bonding temperature and time, because these parameters affect the concentration of diffusing of the components at the interface and, therefore, the nature of the resulting interface.

2. *In a second case*, MMC/AZ91E combinations were produced by solid-state diffusion joining of AlN-AZ91E composite joint to Ti metal at 600 °C for 5 h under argon atmosphere. It is possible to observe a cross section of the interface produced in this joint in Fig. 5.11 [41].

Fig. 5.11 shows the following characteristics: a homogeneous bonding of Ti to AlN-AZ91E MMC material at 600 °C for 5 h by solid diffusion bonding, joint combination formed through the diffusion of Ti with the Mg alloy in the matrix of the MMC, a continuous and homogeneous diffusion zone free of porosity and thermal cracks, and bonding of Ti and AlN-AZ91E MMC material which occurred at relatively low temperature compared with the fusion point of Ti (1560 °C); however, the high affinity between Ti and Mg of the MMC matrix during the process promoted interfacial interactions and bonding between the materials.

The EPMA-EDS carried out in these samples shows an overview of the different components across the Ti/AlN-AZ91E interface using element line scan analyses, and the results are illustrated in Fig. 5.12.

The elements analyzed were Ti, Al, Mg, Mn, Si, and O. The analysis was performed starting on the MMC side of the sample through the bonding line up to

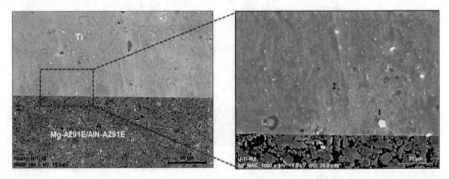

Fig. 5.11 SEM image of the cross-sectional bonding of Ti/AlN-AZ91E combinations joined at 600 °C for 5 h [41]

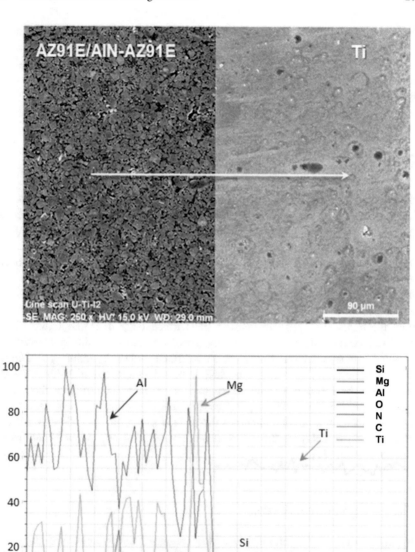

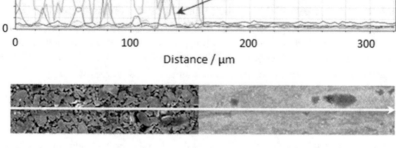

Fig. 5.12 Electron probe line microanalysis through the cross section of Ti/AlN-AZ91E combinations joined at 600 °C for 5 h [41]

the pure Ti, observing the changes of the Mg and Al profiles inside the MMC corresponding to the analyses inside the matrix and AlN reinforcement, respectively; there are some points where the concentration of Si increases inside the MMC, which is produced by the effect of the temperature and time during the bonding process.

5.2.5 Liquid-State Diffusion Bonding

Liquid-phase joining, also called *indirect bonding*, is a technique that involves the liquid formation in the interaction between materials, but also it is considered the most common technique to achieve high integrity joints using a wide range of intermediate bonding materials such as organic adhesives and glasses and processes such as welding, soldering, and brazing. Sometimes it is necessary to use pretreatment or coating on the MMC material surface to improve wetting [42, 43]. A popular technique adopted to enhance wetting is to add suitable alloying elements to the melt. Alloying additions can promote wetting with a solid surface by three mechanisms: (1) by decreasing the surface tension of the liquid due to adsorption onto the surface of the liquid, (2) by decreasing the solid-liquid interfacial tension due to segregation of solute to the interface, and (3) by inducing a chemical reaction at the solid-liquid interface which in turn decreases the solid-liquid interfacial tension by forming a stable compound at the interface. Another method for improving wettability is the application of a metal coating to the surface of the ceramic which, in principle, increases the surface energy of the solid and acts as a diffusion barrier delaying the onset of reaction. Nickel and copper coatings are commonly used [44].

5.2.5.1 Conventional Brazing

Brazing is one of the joining processes associated with liquid filler formation between the base materials. This process has been defined as joining technique that takes place above 450 °C which is based on melting and solidification of an interlayer filler metal or filler alloy which flows by capillary forces between the materials at the joint and whose melting temperature is lower than the solidus temperature of the base materials. Compared with other joining technologies, brazing presents some advantages that can be summarized as follows [45]:

- Low cost at industrial scale due to the equipment which requires relatively little capital.
- Complex and large geometries can be jointed.
- Hybrid materials with different thicknesses can be bonded in a single-step operation.

However, the main disadvantage found in the brazing process is the melting point of fillers or filler alloys which in turn leads to limiting the service temperature of assembly [45, 46]. On the other hand, wettability and contact angle constitute two of the biggest problems during joining due to most of the metal or light alloys not wetting ceramics, and in this technique, the wettability is the capability of solids to build interfaces with liquids; in other words, it is the ability of a liquid to spread over a solid surface. Wettability describes the extent of intimate contact between a liquid and a solid; but it does not represent an index of the strength of the interface. The wettability of a given solid-liquid couple can be measured by considering the equilibrium forces in a system consisting of a drop resting on a flat solid surface in a given atmosphere; this method for evaluating wettability is called the sessile drop method; however, this method is an oversimplification of true and complex phenomena. The contact angle of a liquid on the solid surface is the parameter used to measure the degree of wetting; $\theta > 90°$ means non-wetting, $\theta = 0°$ means perfect wetting, and $\theta < 90°$ indicates partial wetting. It is, however, ordinary practice to say that a liquid wets a solid when $\theta < 90°$.

The shape of the drop results from the balance between the surface force and the interfacial forces that are trying to minimize the surface free energy of the system. Under thermodynamic equilibrium and steady-state conditions, the contact angle is related to the three tensions of the interfaces solid/gas, solid-liquid, and liquid/gas. High temperatures and extended contact times usually promote chemical reactions enhancing or inducing wetting. Leon et al. [47] studied the effect of time at temperature of 900 °C on the contact angles of several Al alloys with TiC ceramic in vacuum and argon; their results were showed in Figs. 2.46 and 2.47.

The criteria for selection of brazing alloys are that they must wet or coat the ceramic, they must form a chemical bond at the interfaces resulting in a strong joint, and they should cause minimal degradation of the base material or materials. Successful brazing alloys produce bonds that are, strong, reliable, and relatively inexpensive to manufacture. As is the case for any joining process, there are also some important constraints and concerns, many of which are the direct consequence of the presence and action of the reactive metal.

5.2.5.2   Reactive Brazing

During brazing process for MMC or ceramic to metal or alloy joining, it is essential that the molten filler metal wets the MMC material. To promote wetting, the surface of the MMC could be previously metalized with active elements such as Ti, Hf, Ni, Nb, Ta, Cr, etc.. Do-Nacimento et al. [48] used the mechanical metallization with Ti as an alternative route to deposit active metallic film on Si_3N_4 ceramic surface prior to brazing process. They argued that addition of small amount of Ti into convectional alloys favored the development of the join. They concluded that the Ti films mechanically deposited onto Si_3N_4 surfaces reacted with the filler alloys, improving their activity and wetting, where there is no evidence of thermal cracking. As a different alternative, the filler metal compositions should contain an active element which in turn leads to the development of the new commercial filler alloys for high-temperature service.

On the other hand, Paulasto et al. [49] concluded that the active element can destabilize the ionic-covalent bond of the ceramic and form an intermediate reaction layer which in turn leads to achievement of chemical bonding between ceramic and metal, as well the content of active element in filler must be reasonable; otherwise, the brittleness of the joint is increased due to the excessive reaction which is associated with liberation of Si atoms to form brittle silicides. Several factors that could weaken the interface of the joint have been studied. However, evidence has shown that when different amounts of additives are used in Si_3N_4 ceramics, these have effects on rate of decomposition in Si_3N_4 during ceramic to metal joining process as showed in Ceja et al. [50] in $Si_3N_4/Cu\text{-}Zn/Nb$ bonding at 1000 °C for 40 min (Fig. 5.13), while with 4 wt% additives used in Si_3N_4 ceramics, it is possible to observe Si-based components near the bonding line between reaction interface and Nb (Fig. 5.13a). But when using the same joining conditions with Si_3N_4

Fig. 5.13 Cross section of the $Si_3N_4/Cu\text{-}Zn/Nb$ interface for sample brazed at 1000 °C for 40 min in argon: (**a**) M2 specimen, having 4 wt% total additives, and (**b**) M3 specimen, having 8 wt% total additives [50]

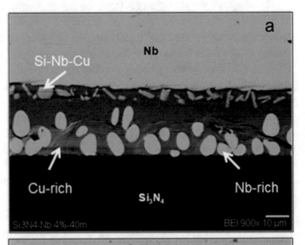

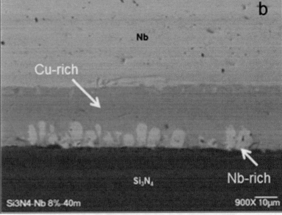

ceramics having 8 wt% additives, no silicide formations on interface were detected (Fig. 5.13b). It is evident that in our case, Si-based component formation should be avoided, if the amount of additives used is 8 wt%; therefore, the amounts of additives have significant effects on the silicon nitride decomposition which in turn leads to silicide formations on interface in Si_3N_4 ceramic to metal joining.

On the other hand, Huang et al. [51, 52] made an investigation during reactive diffusion bonding of SiC_p/Al6063 metal matrix composites using mixed Al-Si, Al-Si-SiC, and Al-Si-W powders as interlayer. The results showed that using mixed Al-Si-SiC powder as insert layers, SiC_p/6063 MMC can be reactive diffusion bonded by a composite joint; however they observed low shear strength of the reactive joints due to the segregation of the SiC forming a number of porous zones in the joint layer. The Ti added in the insert layer obviously improved the joint strength of the mixed Al-Si-SiC powder. Nearly all the W added into the insert layer of Al-Si-W reacts with Al to form intermetallic WAl12 during bonding. The reaction between the W and Al facilitates the formation of a dense joint of high quality; in addition the WAl12 has a reinforcing effect on the joints, which enables the joints bonded by the insert layer of mixed Al-Si-W powder to have high shear strength.

5.2.6 Partial Transient Liquid-Phase Bonding (PTLPB)

Liquid-state diffusion bonding relies on the formation of a liquid phase at the bond line during an isothermal bonding cycle. This liquid phase then infuses the base material and eventually solidifies as a consequence of continued diffusion of the solute in to the bulk material at constant temperature. Therefore, this process is called transient liquid-phase (TLP) diffusion bonding. Despite the presence of a liquid phase, this process is not a subdivision of brazing or fusion welding as the formation and annihilation of the liquid phase occur at a constant temperature and below the melting point of the base material. The liquid phase in TLP diffusion bonding generally is formed by inserting an interlayer which forms a low melting point phase, e.g., eutectic or peritectic, after preliminary interdiffusion of the interlayer and the base metal at a temperature above the eutectic temperature. Note that the liquid phase could, alternatively, be formed by inserting an interlayer with an appropriate initial composition, e.g., eutectic composition which melts at the bonding temperature. The diffusion rate in the liquid phase enhances dissolution and/or disruption of the oxide layer and so promotes intimate contact between the faying surfaces. Therefore, the presence of a liquid phase reduces the pressure required for TLP diffusion bonding in comparison with solid-state diffusion bonding and may overcome the problem associated with solid-state diffusion bonding of the materials with a stable oxide layer. Achieving high integrity joints with minimal detrimental effects on the parent material in the bond region and also the possibility of joining metal matrix composites (MMC) and dissimilar materials are the most promising features of TLP diffusion bonding.

Nami et al. [53] studied transient liquid-phase diffusion bonding of Al/Mg$_2$Si metal matrix composite; they joined as-cast Al/Mg$_2$Si metal matrix composite by TLP diffusion bonding using Cu interlayer at various bonding temperatures. The metal matrix composite contained 15% Mg$_2$Si and was produced through in situ technique by gravity casting. Specific diffusion bonding process was applied as a low vacuum technique. They found that the microstructure of joints consisted of Al-α, CuA$_{12}$ and Mg$_2$Si, or Al-α and Mg$_2$Si depending on the bonding temperature and duration. The maximum shear strength was achieved when samples were bonded at 580 °C for 120 min. Microhardness and compositional homogeneity of joints across the bonded interface were improved with increasing the bonding time at 560 °C.

The active metal brazing is one of the more flexible techniques and cost-effective in ceramic to metal joining. However, when the joining temperature is elevated (>1000 °C), the brittle compounds' formation into the interface has been observed which in turn leads to changes in the microsructure and thickness of the interface which may lead the joining strength to decrease. This factor has led to exploration of nonconventional methods of joining that would exploit a higher diffusivity transport path at lower temperature such as partial transient liquid-phase bonding (PTLPB). The partial transient liquid-phase bonding technique is a promising joining technology due to the advantage of as much brazing, as diffusion bonding, used. This technique would exploit the liquid film formation and rapid diffusion of the metal with higher melting point within the liquid to facilitate joint formation. Namely, during ceramic-metal joining, the edges of the metal melt directly or form liquid through eutectic reaction, and then the liquid phase accelerates joining due to the movement of atoms in liquid which is quicker allowing the wetting of the ceramics and eliminates the holes at interface which may reduce the joining time and pressure. These factors could be an indicative of a strong joint formed by diffusion of atoms and isothermal solidification.

The complete disappearance of the liquid formed would not be required, only rearrangement of the liquid phase so that the joint region no longer contained a continuous layer of the liquid former. By dispersing the liquid former, melting would have a minimal effect on the high-temperature properties of the joint. This alternative approach opened up a much wider range of alloy systems as interlayer candidates for the future, one in which the liquid former has minimal solubility in the host material [54].

5.3 Diffusion Joining Theory of MMC

Diffusion bonding can be defined as the creation of an intimate bond or joint between two materials by thermally assisted processes occurring in the solid state. In order to understand the mechanisms and driving forces of diffusion bonding, the evolution of the bond microstructure must be appreciated. The bonding process can be viewed as two steps operating in parallel. The first is the transition from two surfaces contacting at their asperities to an intimate interfacial conformity. This must involve the elimination of a large volume of interfacial voids accommodated by mass transfer mechanisms, plastic flow, and diffusion. In parallel with this, but sequential to each individual contact, there must be an adhesion process giving the interphase boundary strength. A third step, with possible destructive consequences, is a subsequent chemical reaction between the metal and ceramic or MMC in contact to form a third phase at the boundary. The driving force for the formation of an interface between materials is the energy decrease of the system resulting from its establishment [55, 56]. The interfacial energy should reach the lowest achievable value as the bond is formed; otherwise further changes that could degrade the stability of the bond may occur under operating conditions.

The mechanisms of diffusion bonding two identical materials and similar surfaces have been studied since the 1960s, and it is now generally accepted that joint formation occurs by collapse of interface voids produced by a number of diffusion and creep mechanisms. The collapse of interfacial voids can be brought about by a number of mechanisms analogous to those occurring in pressure sintering, and these are best grouped in terms of sources and sink for matter and are [57–59]:

1. Surface diffusion from a surface source to a neck
2. Volume diffusion from a surface source to a neck
3. Evaporation from a surface source to condensation at a neck
4. Grain boundary diffusion from an interfacial source to a neck
5. Volume diffusion from an interfacial source to a neck
6. Plastic yielding resulting in deformation of original surface asperities
7. Power law creep

An illustration of the various routes of material transfer is contained in Fig. 5.14. These mechanisms are normally separated in two main stages:

Stage 1: Plastic deformation. The contact area of asperities, though initially small, will rapidly grow until the application load can be supported, which means that the local stress falls below the yield strength of the material.

Stage 2: Diffusion and power law creep. The driving force for mechanisms A, B, and B′ is the difference in surface curvature. Matter is transferred from the point of least curvature (sharp neck of the void at the bond interface) to the point of greatest curvature. Thus, as the voids change from an elliptical to a circular cross section, the rates of these mechanisms will approach zero because the aspect ratio of the voids tends to unity.

In addition to these stages, recrystallization and grain growth may occur during bonding. Interface formation must be accompanied by the collapse and annihilation of voids created at first contact. The driving force for this collapse is identical to what drives diffusion bonding in metals. Void closure results in a net approach of the two surfaces being joined. This allows mechanical work to be done by the bonding pressure. The reduction in void volume is accompanied by a reduction of void surface energy, which is a further driving force. A number of additional competing mechanisms can occur during the bonding of dissimilar materials such as metals and ceramics.

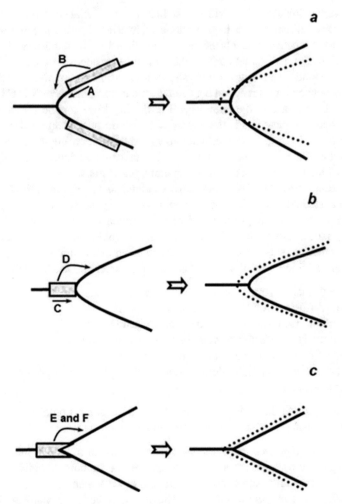

Fig. 5.14 Schematic illustration of material transfers for various mechanisms involved in diffusion bonding: (**a**) transport from surface to an interfacial neck, (**b**) transport from bonding surface to a neck, (**c**) bulk deformation by plastic flow after yield or during creep [59]

In the other hand, when a liquid is present during bonding, the process varies drastically. Some theoretical aspects during transient liquid-phase (TLP) diffusion bonding are present in general shape as follows. Figure 5.15 shows a simple eutectic phase diagram where A represents a pure parent material and B is the diffusing solute (i.e., the originally solid interlayer) with limited solubility in A. Basically, TLP diffusion bonding consists of two major stages: (1) dissolution of the base metal (Fig. 5.15b) and (2) isothermal solidification (Fig. 5.15c).

The dissolution stage can be divided into two hierarchical substages in which filler metal melting is followed by widening of the liquid zone. However, if the

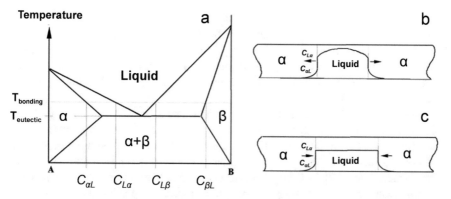

Fig. 5.15 Schematic illustration of (**a**) eutectic phase diagram, where A represents a pure parent material and B is the diffusing solute, and (**b**) dissolution of base metal and homogenization of liquid phase; (**c**) horizontal arrows show the direction of movement of the solid-liquid interface [39, 60]

melting process occurs as a result of interdiffusion of A and B, then melting of the interlayer and widening of the liquid phase may occur simultaneously [39, 60]. According to the phase diagram, equilibrium in the liquid can be established by dissolution of A atoms into the supersaturated Brich liquid to decrease its concentration to CLα. During this stage, homogenization of the liquid phase continues, and the width of the liquid zone increases until the composition profile in the liquid phase levels out, i.e., diffusion in the liquid ceases. The rate of this homogenization is controlled mainly by the diffusion coefficient in the liquid phase, and, therefore, this stage takes a short time to be completed. In the next stage, isothermal solidification occurs as B atoms and starts to diffuse into the solid phase, and the liquid zone shrinks in order to maintain the equilibrium compositions of CLα and CαL at the solid-liquid moving boundaries. The interdiffusion coefficient in the solid phase (α) controls the rate of solidification, and because diffusivity in the solid is low, the annihilation of the liquid phase is very slow compared to the initial rapid dissolution stage for which diffusivity in the liquid controls the rate of the reaction.

5.4   Mechanical Evaluation of MMC Joints

Several problems have been associated with the joint strength influencing the reliability of a joint. Several important defect categories may cause scatter in strength directly; from the microscopic view, the reaction structure caused by wetting or by chemical and physical bond-ability between two faces may be of concern [61]. These factors will reflect the distribution of unjoined or weakly bonded island-like defects on interfaces resulting in substantial reduction in joint strength. Unjoined areas reduce joint strength especially in solid-state joining. From the more macroscopic view, when a reaction layer grows, cracking occurs in the layer, which frequently influences joint strength. Thermal or residual stress in a joint becomes another important factor. The development of residual stresses at the interface when the material is cooled down from the bonding temperature to room temperature is one of the major problems in ceramic-metal joining. These residual stresses reduce the strength of the bonded material and in some cases lead to joint failure during or after the joining process.

The mechanical characterization of a metal-ceramic joint is a complex problem. There are a variety of different properties to be considered in ceramic-metal joints. Depending on the application of the joint, some properties are more important than others. However, the mechanical performance is one of the most important properties for any joint [62]. It is important to distinguish between joints made between *similar* materials, whether they are metals, ceramics, composites, and joints which involve interfaces between *dissimilar* materials, e.g., metal bonded to ceramic or glass or ceramic bonded to glass. In the case of *dissimilar* materials, the engineering compatibility of the two components must be considered. Mismatch of the elastic modulus is a common form of mechanical incompatibility, which leads to stress concentrations and stress discontinuities at the bonded interface between the two materials. Figure 5.16 shows an example in which a normal load is transferred across the interface between two materials with different elastic moduli [63, 64]. The stiffer, higher modulus, component restricts the lateral contraction of the more compliant, lower modulus, component and generates shear stresses at the interface that may lead to debonding.

Thermal expansion mismatch represents a lack of physical compatibility and is a common problem in dissimilar joints. Thermal expansion mismatch leads to the development of thermal stresses which tend to be localized in the joint and reduce its load-carrying capacity, ultimately leading to failure of the component. Poor

Fig. 5.16 Mismatch in the elastic modulus of bonded components results in elastic constraint, which generates shear stresses parallel to the interface under normal loading conditions [59]

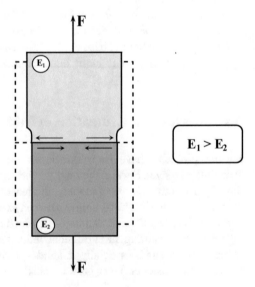

chemical compatibility is commonly associated with undesirable chemical reaction in the region of the joint. Chemical reactivity between the components may lead to undesirable interface reactions and the products of these reactions are often brittle. Reactions accompanied by a volume change generate local stresses, and the mechanical integrity of the joint will be threatened. Thermal expansion mismatch has a great influence not only on the absolute value of strength but also on the reliability of joints.

As the use of advanced ceramic materials for critical structural components increases, reliability in dissimilar joints becomes a critical issue. Ensuring reliable performance in a dissimilar joint means being able to predict with a high degree of certainty whether a component will fail under typical operating conditions. This requires an accurate description of the stresses likely to be encountered in service, as well as knowledge of the mechanical properties of the joint. There are a variety of different properties to be considered in the ceramic-metal joints, e.g., mechanical properties, electrical properties, thermal properties, etc. Depending on the application of the joint, some properties are more important than others. However, the mechanical properties are some of the most important properties for any joint. Joints without any mechanical strength can be regarded as unsuccessful joints. A description of the mechanical behavior of ceramic-metal joints requires the determination of their strength along with the distribution of residual thermal stresses. For the evaluation of the joint properties, it is essential to establish proper testing methods so that effects of the processing can be detected accurately and consistently [62, 65].

In the production of dissimilar joints, the required strength of the joints is an important criterion in the selection of an appropriate joining technique. In contrast to homogeneous materials, such as steels or other metals, the strength of the joint is not a material-specific parameter, which can be found in reference books, but is influenced considerably by the following factors [66, 67]:

- The selection and mechanical properties of the individual joint partner materials
- Differences in the thermal expansion coefficients of the partner materials
- Selection of the joining technique
- Interface reactions between the joining partners
- The design of the joining geometry

By the selection of the materials to be joined, the material-specific constants like Young's modulus and thermal expansion coefficient are determined. The resulting strength of the joint depends to a large extent on the preparation of the materials and the joining technique. Diffusion processes between solid materials can also affect bonding of materials. With this technique, the polished surfaces of the material partners are brought into contact under the application of pressure; they are then heated to a high temperature just below the lowest melting point of the two material partners.

The materials are kept at this temperature for a specific time and then cooled very slowly. In order to determine the success of the respective joining technique and the joining parameters used, and the reproducibility of the joining technique, the strength of the joint must be tested [68]. Strength is one of the critical properties for structural applications. Although there is an *ASTM* standard (tensile test), for the metal/metal joints, most researchers have used their own method for the evaluation of the ceramic-metal joint strength. Because the normal *ASTM* method requires complex shape tensile specimen, an alternative test method needs to be established.

The mechanical characterization of a metal-ceramic joint is a complex problem. Even if there are no residual stresses from the bonding process, the difference in elastic properties on each side of a metal-ceramic joint will lead to interfacial stress concentrations in the absence of a defect. Once a crack is initiated and failure commences, the different strain energy release rates in the metal and ceramic components can lead to deflections of the crack away from the interface. Therefore it may be necessary to adopt a component-based approach to bond strength analysis where many factors of design, and not just the mechanical properties of the bond, are considered.

Several methods have been used to measure joint strength. At present, the most common methods include tensile, bending, or flexural and shear tests, where the stress to fracture the bonded surfaces is used to characterize the joint strength [69, 70]. The schematics of various test methods of joint strength are shown in Fig. 5.17. Tensile tests are generally performed in double-joint specimens, ceramic/metal/ceramic, whereas three-point and four-point bending can be performed in both single and double joints.

Shear tests can only be performed on single joints as a consequence of their intrinsic geometry. The characterization of the interfacial strength by pull-off or shear-off tests has several limitations.

The first one is related to the variety of techniques used by different research groups, making it difficult to establish a mutual comparison of results. The shear test provides an alternative way to assess the mechanical strength of interfaces. Samples are easily produced, but the results are generally lower than those obtained for bend and tensile tests.

The selection of an appropriate method for measuring the bond strength is dictated by the purpose of testing, but the bonding process and parameters affecting the mechanical quality of the bond can be monitored by both fracture mechanics and conventional testing methods. The bond strength values obtained also depend on the testing technique chosen. Bend test values are generally higher than tensile test values for joints and for brittle ceramic materials. The shear test is one of the simplest methods. However, the shear stress at the interface is not simple shear, and it always contains a component of tensile stress that originates from a bending moment, which cannot be neglected. The influence of a slight change of the push position and the fixing condition on the stress distribution is very important. Therefore, the shear test is not recommended for the common evaluation method. Bending and tensile test have almost the same stress distributions as those derived from analytical equations. However, the elastic constant mismatch between ceramics and metal induces inhomogeneity in stress distribution [71]. The tensile test requires a careful preparation of the test specimen and a strict alignment of the load train.

These difficulties in testing will influence the reproducibility of the strength measurements. On the contrary, bending tests have greater flexibility compared to

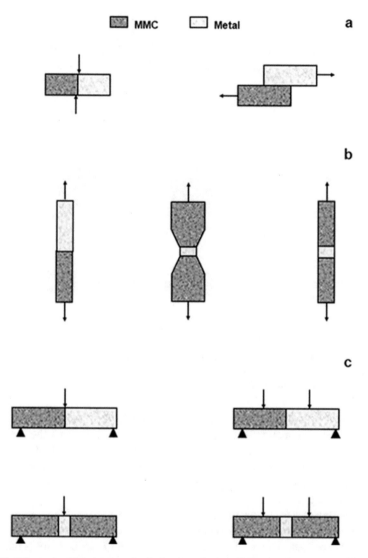

Fig. 5.17 Schematics of various mechanical test methods performed on ceramic-metal joints: (**a**) shear test, (**b**) tensile test, (**c**) bend test on single and double joint [59]

the tensile test. However, in the case where plastic deformation occurs in the metal, the analytical equation for bending stress becomes complicated. Figure 5.18 shows a schematic comparison of the tensile stress distribution for three-point, four-point, and uniaxial tensile test specimens [72]. In the case of three-point bending, the peak stress occurs only along a single line on the surface of the test bar opposite the point of loading. The tensile stress decreases linearly along the length of the bar into the thickness of the bar, reaching zero at each bottom support and at the neutral axis,

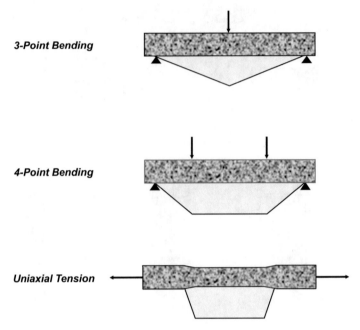

Fig. 5.18 Comparison of the tensile stress distributions for three-point, four-point, and uniaxial tensile test specimens. Shaded area represents the tensile stress, ranging from zero at each support of the bend specimens to maximum at midspan and being uniformly maximum along the whole gauge length of the tensile specimen [59]

respectively. The probability of the largest flaw in the specimen is very low. Therefore, the specimen will fracture at either a flaw smaller than the largest flaw or a region of lower stress. Four-point bend testing results in lower strength values for a given ceramic material than does three-point bending. The peak for the stress distribution in a four-point bend specimen is present over the area of the tensile face between the load points.

The tensile stress decreases linearly from the surface to zero at the neutral axis and from the load point to zero at the bottom supports. The area and volume under peak tensile stress or near peak tensile stress is much greater for four-point bending than for three-point bending, and thus the probability of a larger flaw being exposed to high stress is increased. As a result, the modulus of rupture or bend strength measured in four-point is lower than that measured in three-point. Uniaxial tensile strength results in lower strength values for a given ceramic than does bend testing. Figure 5.18 illustrates that in the case of uniaxial tension, the complete volume of the gauge section of a tensile test specimen is exposed to the peak tensile stress. Therefore, the largest flaw in this volume will be the critical flaw and will result in fracture.

The interfacial strength of dissimilar joints materials is generally determined by bend testing, also referred to as flexure testing. The test specimen can have a circular, square, or rectangular cross section and is uniform along the complete length. As

Fig. 5.19 Schematic setup of the mechanical arrangement used to evaluate the four-point bend strength of joints [74]

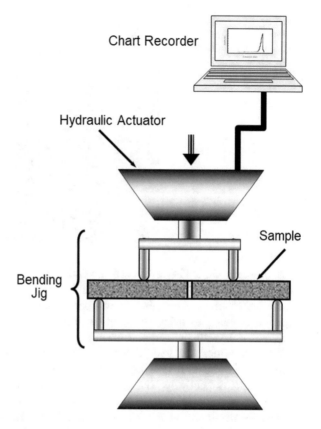

shown in Fig. 5.19, the test specimen is supported near the ends, and the load is applied either at the center, for three-point loading, or at two positions for four-point loading. The final goal for joining research will be in establishing a technique producing a tightly bound interface by eliminating defects and by accommodating thermal stress. The specimen is carefully placed in the bottom part of the jig, with the interface plane parallel to the plane of vertical displacement of the plunger. The load is applied at a low vertical speed until the fracture of the specimen [73]. The bend strength is defined as the maximum tensile stress at failure and is often referred to as the modulus of rupture (*MOR*). The bend strength for a circular test specimen can be calculated using the general flexure stress formula described in Eq. (5.1) [72]:

$$X = \sigma_{4-\text{pt}} = 16\,F \cdot d/\pi D^3 \tag{5.1}$$

where F is the load at fracture, d is the distance between the outer and inner span of the four-point bend jig, and D is the diameter of the specimen. For each set of experimental conditions studied, temperature and time, an average of at least five samples must be used to determine the bending strength for each joining condition.

Fig. 5.20 Line analysis
through the interface
obtained in a WC-Co/Cu-
Zn/Ni sample joined at
980 °C for 15 min [74]

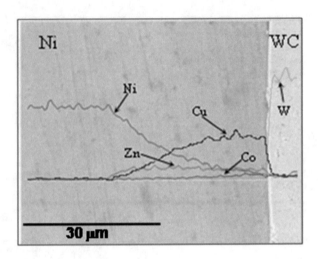

A previous work [74] evaluated the interfacial strength of WC-Co/Zn-Cu/Ni brazing samples produced at 980 °C at different holding times under argon atmosphere. Microstructural examination was performed on polished cross sections of the experimental couples using scanning electron microscopy and microanalysis. An overview of the different components in the interface was obtained in a WC-Co/Zn-Cu/Ni sample joined at 980 °C for 15 min by line analysis using electron probe microanalysis. The results are illustrated in Fig. 5.20 where the Ni and WC are on the left and right, respectively. The scan line was chosen to start on the Ni side of the sample through the interface, Cu-Zn, finishing on the WC side. The Ni signal reached its maximum at the Ni-Cu brazes. Interdiffusion of Cu-Ni and Cu-Co could be observed. The microanalysis profile indicates the presence and even distribution of Zn concentration. In the region corresponding to diffusion zone, high levels of Cu and Zn were observed; however, evaporations of Zn during bonding occurred.

The interfacial strength of WC-Co/Cu-Zn/Ni joints was determined by a four-point bending test using a universal testing system with a 25 KN load cell and a bending jig. The load was applied a vertical speed of 0.5 mm/min until the applied load resulted in fracture of the specimen. For each set of experimental conditions studied, temperature and time, an average of at least four samples of 50 mm in length and 6.35 mm in diameter were used to determine the bend strength; the results obtained are shown in Fig. 5.21. It can be observed that the strength of the joint increased from a value of 233 MPa and reached a maximum value of 255 MPa, when the time was increased from 5 to 15 min, respectively, and decreased beyond this time. This improvement was attributed to the increase of interface reaction and formation of a strong chemical bridge between the two materials. On the other hand, the thickness of the reaction zone increases with time and may dominate in the final strength.

Reaction products are generally brittle, and as the thickness of these phases increases, the joint strength, at first, rises due to the creation of a strong, integral

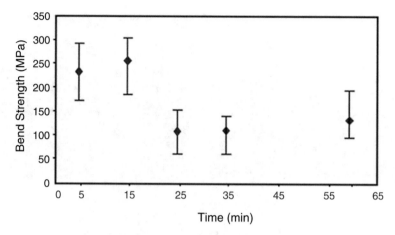

Fig. 5.21 Modulus of rupture for WC-Co/Cu-Zn/Ni samples joined at 980 °C [74]

Fig. 5.22 Bend test sample
diffusion joining a 980 °C
and 15 min before
testing [74]

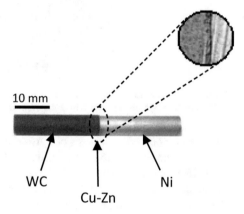

bond and then reaches a maximum at a certain thickness and then decreases as the
interface continues to grow. The strongest joint was obtained for joining conditions of
980 °C and 15 min, with a resulting average bending strength of 255 MPa; however,
joint strengths greater than 100 MPa in average were produced at 980 °C and times
vary from 5 to 60 min. An example of a test sample is show in Fig. 5.22.

On the other hand, Castro-Sanchez et al. [75] evaluated the interfacial strength,
using a shear test method, of WC-Co/Zn-Cu/Inconel 600 brazing samples produced
at 1000 °C at different holding times under argon atmosphere. Figure 5.23 shows a
cross section of the interface observed in WC-Co/Zn-Cu/Inconel 600 brazing samples produced at 1000 °C for 35 min. The interface was obtained by secondary
(Fig. 5.23a) and backscattering (Fig. 5.23c) electron image. It can be observed
clearly from the concentration of Cr in the dark layer (Fig. 5.23b) and points inside
the Inconel 600, confirming the diffusion of Cr during bonding process, as well as
some point close to the Inconel 600 bonding lines.

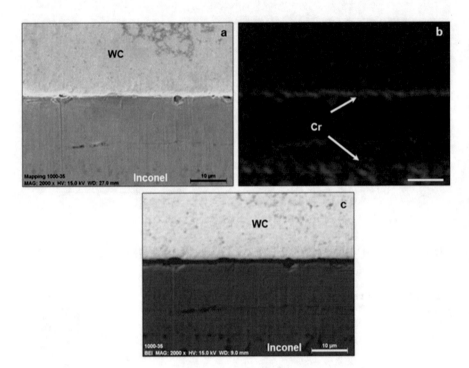

Fig. 5.23 Cross section of the interface in a WC-Co/Cu-Zn/Inconel 600 sample joined at 1000 °C for 35 min: (**a**) SEM image, (**b**) Cr mapping, and (**c**) backscattering electron image [75]

Electron probe microanalysis performed on these samples confirmed that Co, Cu, Zn, Ni, Cr, and Fe were the main diffusing elements into the interface. An overview of the different components in the interface was obtained in a WC-Co/Zn-Cu/Inconel 600 sample joined at 1000 °C for 35 min by line analysis using electron probe microanalysis. The results are illustrated in Fig. 5.24 where the WC-Co and Inconel 600 are on the left and right, respectively. The scan line was chosen to start on the WC side of the sample through the interface, Cu-Zn, finishing on the Inconel 600 side. The Ni, Cr, and Fe signals reached its maximum at the Inconel-Cu boundary. Interdiffusion of Inconel-Cu and Co-Cu throughout the interface could be observed. It is an important remark how the concentration of Cr increases inside the dark layer and the points revealed by backscattering electron image close to the WC-Co and Inconel 600, respectively. In the region corresponding to diffusion zone, high levels of Cu and Zn were observed; however, evaporations of Zn during bonding could happen as observed by Zhang et al. [76].

In order to establish a mechanical evaluation, the interfacial strength of WC-Co/ Cu-Zn/Inconel 600 joints was determined by shear test using a universal testing system with a 25 kN load cell. The load was applied at a constant vertical speed of 0.1 mm/min. The schematic illustration of the shear test equipment is shown Fig. 5.25 [77]. The results obtained for WC-Co/Cu-Zn/Inconel 600 joints produced at 1000 °C are shown in Fig. 5.26. The error bars correspond to plus or minus the

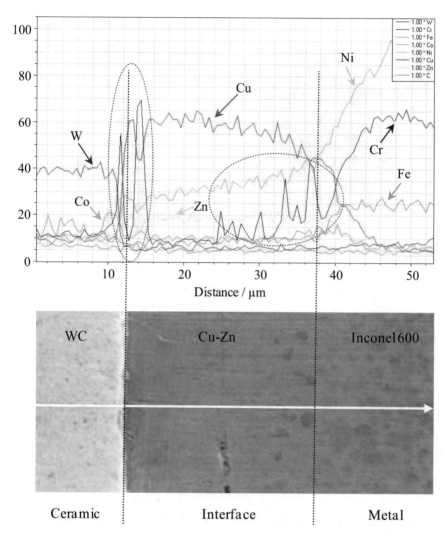

Fig. 5.24 Line analyses through the interface in a WC/Cu-Zn/Inconel 600 sample joined at 1000 °C for 35 min [75]

standard deviation for the average joint strength of at least three samples of 20 mm in length and 6.35 mm in diameter for each set of experimental conditions.

It can be observed in Fig. 5.26 that the strength of the joint increased from a value of 35 MPa and reached a maximum value of 44 MPa, when the time was increased from 15 to 25 min, respectively, and decreased beyond this time (35 min) until a value of 29 MPa. This improvement was attributed to the increase of interface diffusion and formation of a strong chemical bridge between the two materials. On the other hand, the thickness of the reaction zone increases with time and may dominate in the final strength. Reaction products are generally brittle, and as the

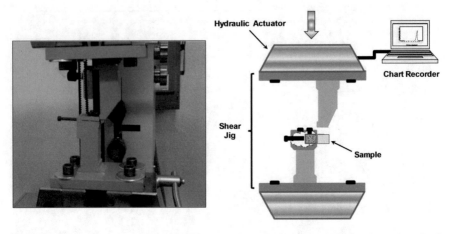

Fig. 5.25 Schematic setup of the mechanical arrangement used to evaluate the shear strength of joints [77]

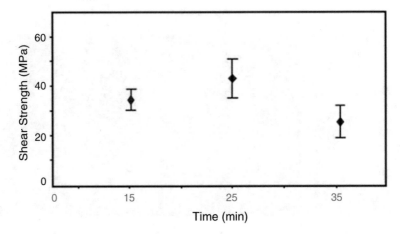

Fig. 5.26 Shear strength in function of time for WC-Co/Cu-Zn/Inconel 600 sample joined at 1000 °C [75]

thickness of these phases increases, the joint strength, at first, rises due to the creation of a strong, integral bond and then reaches a maximum and then decreases as the interface continues to grow, as the presence of the dark layer close to WC-Co bonding line showed in Fig. 5.23c.

The effect of a reaction layer on the interface strength depends on a number of factors such as the mechanical properties of the reaction layer, its thickness, and morphology. The choice of suitable conditions to prepare ceramic-metal joints requires knowledge concerning the mechanism of reaction between the materials and the evolution of the interface. Therefore, the reaction layer must be controlled in order to ensure good joint strength. The main part of the fracture surface occurred along the Inconel 600 diffusion zone interface and probably initiated at the edge of the sample.

In summary, the choice of suitable conditions to prepare ceramic-metal joints requires knowledge about the mechanism of reaction between the materials and the evolution of the interface. The maximum value of 44 MPa for the shear strength of the joint was obtained for samples diffusion bonded at 1000 °C for 25 min. This strength is of the same order as the shear strength reported for Kenevisi and Khoie [78] to joining Ti-6Al-4 V to Al7075 (30 MPa) and Samavatian et al. [79] to joining Ti-6Al-4 V to Al2024 using a cu-Zn alloy (37 MPa).

References

1. Fernandez P, Martínez V, Valencia M et al (2006) Applications of metal matrix composites in electric and electronic industries. Dyna Rev Fac Nac Minas 73(149):1–8
2. Gay D (2015) Composite materials: design and applications, 3rd edn. CRC Press/Taylor & Francis, Boca Raton
3. Miracle DB (2005) Metal matrix composites – from science to technological significance. Compos Sci Technol 65:2526–2540
4. Rajeshwar K, De Tacconi NR, Chenthamarakshan CR (2001) Semiconductor – based composite materials: preparation, properties and performance. Chem Mater 9(13):2765–2782
5. Rawal S (2001) Metal matrix composites for space applications. JOM 53(4):14–17
6. Kainer KU (2006) Chapter 1. Basics of metal matrix composites. In: Metal matrix composites: custom-made materials for automotive and aerospace engineering. Wiley, Weinheim, pp 2–55
7. Kudela S (2003) Magnesium-lithium matrix composites-an overview. Int J Mater Prod Technol 18(1–3):91–115
8. Kainer KU (2006) Metal matrix composites: custom-made materials for automotive and aerospace engineering. Wiley, Weinheim, pp 1–54
9. Prater T (2011) Solid-state joining of metal matrix composites: a survey of challenges and potential solutions. Mater Manuf Proc 26(4):1–23
10. Zhang XP, Quan GF, Wei W (1999) Preliminary investigation on joining performance of SiC_p-reinforced aluminium metal matrix composite (Al/SiC_p–MMC) by vacuum brazing. Compos Part A 30:823–827
11. Lemus-Ruiz J, Ceja-Cárdenas L, Bedolla-Becerril E et al (2011) Chapter 10. Production, characterization, and mechanical evaluation of dissimilar metal/ceramic joints. In: Cuppoletti J (ed) Nanocomposites with unique properties and applications in medicine and industry. InTech, Rijeka, pp 205–224
12. Santella ML (1992) A review of techniques for joining advanced ceramics. Ceram Bull 71 (6):947–954
13. Loehman RE, Tomsia AP (1988) Joining of ceramics. Ceram Bull 67(2):375–380
14. Nicholas MG (1989) Joining structural ceramics. In: Peteves SD (ed) Designing interfaces for technological applications. Elsevier, Amsterdam, pp 49–76
15. Okamoto T (1990) Interfacial structure of metal-ceramic joints. ISIJ Int 30(12):1033–1034
16. Thomas WM, Threadgill PL, Nicholas ED (1999) Feasibility of friction stir welding steel. Sci Technol Weld Join 4(6):365–372
17. Sathiya P, Aravindan S, Noorul Haq A (2007) Effect of friction welding parameters on mechanical and metallurgical properties of ferritic stainless steel. Int J Adv Manuf Technol 31:1076–1082
18. Maalekian M (2007) A friction welding – critical assessment of literature. Sci Technol Weld Join 12(8):738–759
19. Mishra RS, Ma ZY (2005) Friction stir welding and processing. Mater Sci Eng Rep 50:1–78

20. Thomas WM, Nicholas ED (1997) Friction stir welding for the transportation industries. Mater Des 18(4/6):269–273
21. Prater T (2014) Friction stir welding of metal matrix composites for use in aerospace structures. Acta Astronaut 93:366–373
22. Meshram SD, Mohandas T, Madhusudhan G (2007) Friction welding of dissimilar pure metals. J Mater Proc Technol 184:330–337
23. Uday MB, Ahmad-Fauzi MN, Mohd Noor A et al (2016) Chapter 8. Current issues and problems in the joining of ceramic to metal. In: Joining technologies. InTech, Rijeka, pp 159–193
24. Hupston G, Jacobson DM (2004) Principles of soldering. ASM International, Ohio
25. Koshiishi F (2016) Welding duplex stainless steel. Kobelco Weld Today 19:1–10
26. Srivastava AK, Sharma A (2017) Advances in joining and welding technologies for automotive and electronic applications. Am J Mater Eng Technol 5(1):7–13
27. Raghavendra DR, Sethuram D, Raghupathy VP (2015) Comparison of friction welding technologies. Int J Innov Sci Eng Technol 2(12):492–499
28. Urena A, Escalera MD, Gil L (2000) Influence of interface reactions on fracture mechanisms in TIG arc-welded aluminium matrix composites. Compos Sci Technol 60:613–622
29. Lienert TJ, Brandon ED, Lipolds JC (1993) Laser and electron beam welding of SiC_p/A-356 MMC. Scr Metall Mater 28:1341–1346
30. Garcia R, Lopez VH, Bedolla E et al (2002) MIG welding process with indirect electric arc. J Mater Sci Lett 21(24):1965–1967
31. Garciia R, Lopez VH, Bedolla E (2003) A comparative study of the MIG welding of Al/TiC composites using direct and indirect electric arc processes. J Mater Sci 38:2771–2779
32. Garciia R, Lopez VH, Bedolla B (2007) Welding of aluminium by the MIG process with indirect electric arc (MIG-IEA). J Mater Sci 42:7956–7963
33. Suganuma K, Okamoto T, Koizumi M et al (1985) Method for preventing thermal expansion mismatch effect in ceramic-metal joining. J Mater Sci Lett 4:648–650
34. Dunford DV, Wisbey A (1993) Diffusion bonding of advanced aerospace metallics. Mater Res Soc Symp Proc 314:39–50
35. Peteves SD, Nicholas MG (1991) Materials factors affecting joining of silicon nitride ceramics. In: Kumar P, Greenhut VA (eds) Metal-ceramic joining. The Minerals, Metals & Materials Society, Warrendale, pp 43–65
36. Barnes TA, Pashby IR (2000) Joining techniques for aluminium space frames used in automobiles part I-D solid and liquid phase welding. J Mater Proc Technol 99:62–71
37. Surappa MK (2003) Aluminium matrix composites: challenges and opportunities. Sadhana 28 (1–3):319–334
38. Nami H, Halvaee A, Adgi H et al (2010) Investigation on microstructure and mechanical properties of diffusion bonded Al/Mg_2Si metal matrix composite using copper interlayer. J Mater Proc Technol 210:1282–1289
39. Shirzadi AA, Assadi H, Wallach ER (2001) Interface evolution and bond strength when diffusion bonding materials with stable oxide films. Surf Interface Anal 31:609–618
40. Bedolla E, Lemus-Ruiz J, Contreras A (2012) Synthesis and characterization of Mg-AZ91/AlN composites. Mater Des 38:91–98
41. Ortega-Silva E (2016) Producción y caracterización de ensambles híbridos de un material compuesto AlN/MgAZ91E. Dissertation of Master Thesis, Instituto de Investigación en Metalurgia y Materiales, UMSNH, Morelia, México
42. Yong Z, Di F, Zhi-Yomg H et al (2006) Progress in joining ceramics to metals. J Iron Steel Res Int 13(2):1–5
43. Martinelli AE, Hadian AM, Drew RAL (1997) A review on joining non-oxide ceramics to metals. J Can Ceram Soc 66(4):276–284
44. Yokokawa H, Sakai N, Kawada T (1991) Chemical potential diagram of Al-Ti-C system: Al_4C_3 formation on TiC formed in Al-Ti liquids containing carbon. Metall Mater Trans A 22:3075–3076

45. Schwartz MM (1990) Ceramic joining. ASM International, Ohio, pp 99–103
46. Nicholas MG (1998) Joining processes, introduction to brazing and diffusion bonding. Springer, New York, pp 22–24
47. Contreras A, Lopez VH, Leon CA et al (2001) The relation between wetting and infiltration behavior in the Al-1010/TiC and Al-2024/TiC systems. Adv Technol Mater Mater Process J 3:27–34
48. Do Nacimento RM, Martinelli AE, Buschinelli AJA et al (2005) Microstructure of brazed joints between mechanically metallized Si_3N_4 and stainless steel. J Mater Sci 40(17):4549–4556
49. Paulasto M, Kivilahti JK (1995) Formation of interfacial microstructure in brazing of Si_3N_4 with Ti-activated Ag-Cu filler alloys. Scr Metall Mater 32(8):1209–1214
50. Ceja-Cárdenas L, Lemus-Ruiz J, De la Torre SD et al (2013) Interfacial behavior in the brazing of silicon nitride joint using an Nb-foil interlayer. J Mater Process Technol 213(3):411–417
51. Huang JH, Dong YL, Wan Y et al (2008) Reactive diffusion bonding of SiC_p/Al composites by insert layers of mixed powders. Mater Sci Technol 21(10):1217–1221
52. Huang JH, Dong YL, Wan Y et al (2007) Investigation on reactive diffusion bonding of SiC_p/6063 MMC by using mixed powders as interlayers. J Mater Process Technol 190:312–316
53. Nami H, Halvaee A, Adgi H (2011) Transient liquid phase diffusion bonding of Al/Mg_2Si metal matrix composite. Mater Des 32:3957–3965
54. Sugar JD, McKown JT, Akashi T et al (2006) Transient-liquid-phase and liquid-film-assisted joining of ceramics. J Eur Ceram Soc 26:363–372
55. Derby B, Wallach ER (1982) Theoretical model for diffusion bonding. Metal Sci 16(1):49–56
56. Locatelli MR, Dalgleish BJ, Nakashima K et al (1997) New approaches to joining ceramics for high-temperature applications. Ceram Int 23:313–322
57. Chen IW, Argon AS (1981) Diffusive growth of grain-boundary cavities. Acta Metall 29:1759–1768
58. Almond EA, Cottenden AM, Gee MG (1983) Metallurgy of interfaces in hard-metal/metal diffusion bonds. Metals Sci 17:153–158
59. Lemus-Ruiz J (2000) Diffusion bonding of silicon nitride to titanium. PhD Thesis, McGill University, Canada
60. Shirzadi AA, Wallach ER (2004) New method to diffusion bond supralloys. Sci Technol Weld Join 9(1):37–40
61. Suganuma K (1993) Reliability factors in ceramic/metal joining. Mater Res Soc Symp Proc 314:51–60
62. Anderson RM (1989) Testing advanced ceramics. Adv Mater Proc 3:31–36
63. Brandon D, Kaplan WD (1997) Joining processes – an introduction. Wiley, New York
64. Suganuma K, Okamoto T, Koizumi M (1984) Effect of interlayers in ceramic-metal joints with thermal expansion mismatches. J Am Ceram Soc 67:256–257
65. Mizuhara H, Huebel E, Oyama T (1989) High-reliability joining of ceramic to metal. Ceram Bull 68(9):1591–1599
66. Mülheim MT (1994) Bending test for active brazed metal/ceramic joints–a round Robin. Ceram Forum Int 71(7):406–411
67. Lee WC (1997) Strength of Si_3N_4/Ni-Cr-Fe alloy joints with test methods: shear, tension, three-point and four-point bending. J Mater Sci 32:6657–6660
68. Quinn GD, Morrell R (1991) Design data for engineering ceramics: a review of the flexure test. J Am Ceram Soc 74(9):2037–2066
69. Cam G, Bohm K-H, Mullauer J et al (1996) The fracture behavior of diffusion bonding duplex gamma TiAl. JOM 48:66–68
70. Quinn GD (1991) Strength and proof testing. Engineered materials handbook 4, ASM International, Ohio, pp. 585–598
71. Emsley J (1991) The key to the elements. Clarendom Press, Oxford
72. Richerson DW (1992) Modern ceramic engineering, 2nd edn. Marcel Dekker, New York
73. Lemus-Ruiz J, Aguilar-Reyes EA (2004) Mechanical properties of silicon nitride joints using a Ti-foil interlayer. Mater Lett 58(19):2340–2344

74. Lemus-Ruiz J, Ceja-Cárdenas L, Salas-Villaseñor AL et al (2009) Mechanical evaluation of tungsten carbide/nickel joints produced by direct diffusion bonding and using a Cu-Zn alloy. In: International brazing and soldering conference. American Welding Society, Miami, pp 206–212
75. Castro-Sánchez G, Otero-Vázquez CI, Lemus-Ruiz J (2017) Fabrication and evaluation of hybrid components of WC/Inconel 600 by liquid state diffusion bonding. J Mater Sci Eng Adv Technol 16(1):1–16
76. Zhang J, Fang HY, Zhou Y et al (2003) Effect of bonding condition on microstructure and properties of the Si_3N_4/Si_3N_4 joint brazed using Cu-Zn-Ti filler alloy. Key Eng Mater 249:255–260
77. Lemus-Ruiz J, Verduzco JA, González-Sánchez J et al (2015) Characterization, shear strength and corrosion resistance of self-joining AISI 304 using a Ni-Fe-Cr-Si metallic glass foil. J Mater Process Technol 223:16–21
78. Kenevisi MS, Khoie SMM (2012) A study on the effect of bonding time on the properties of Al7075 to Ti–6Al–4V diffusion bonded joint. Mater Lett 76:144–146
79. Samavatian M, Halvaee A, Amadeh AA et al (2015) Transient liquid phase bonding of Al 2024 to Ti–6Al–4V alloy using Cu–Zn interlayer. Trans Nonferrous Met Soc China 25:770–775

Chapter 6
Corrosion of Composites

6.1 Corrosion Behavior of Al-2024/TiC, Al-Mg$_x$/TiC, and Al-Cu$_x$/TiC Composites

The corrosion resistance of commercial aluminum alloy (2024) and binary Al-Cu$_x$ and Al-Mg$_x$ alloys reinforced with TiC particles using a pressureless infiltration method has been evaluated in 3.5% NaCl solution using potentiodynamic polarization curves and linear polarization resistance measurements [1].

Galvanic corrosion between the reinforcement and the metal matrix governs the corrosion behavior of many MMCs [2]. Other factors such as residual contaminants of MMC processing and the formation of interphases between reinforcement and matrix can also have pronounced effects on MMC corrosion behavior. The lack of inherent resistance to corrosion of some MMCs requires that they be coated with organic or inorganic coatings for protection.

In general, the composites are more susceptible to corrosion attack than the matrix alloy. Corrosion tends to be localized with galvanic couples [3]. Aluminum with additions of Mg and Cu was used like matrix in the fabrication of composites with applications where a combination of corrosion resistance, low density, and high mechanical performance is requested, such as in automotive and aerospace industry.

Magnesium alloys have high strength/weight ratios, but poor corrosion resistance affects the viability of increased magnesium usage in aerospace and other applications. Corrosion control measures developed for Mg alloys include the development of high purity or new alloys, rapid solidification, coatings, and surface modification [4].

Gusieva et al. [5] review the influence of alloying on the corrosion of Mg alloys, with particular emphasis on the underlying electrochemical kinetics that dictate the ultimate corrosion rate.

Melchers [6] studied the evolution of corrosion loss and maximum pit depth of copper and copper alloys exposed for long periods of time in natural and industrial.

© Springer Nature Switzerland AG 2018
A. Contreras Cuevas et al., *Metal Matrix Composites*,
https://doi.org/10.1007/978-3-319-91854-9_6

He found bimodal behavior signals that change from mainly cathodic oxygen reduction to a subsequent transitory corrosion process that may be modeled as involving pitting under copper corrosion products. Also discussed are the possible reasons for some data sets showing decreasing maximum pit depths with increasing exposure time.

The reinforcement based on TiC particles is attractive for the aluminum matrix because of its good wettability [7, 8] which can result in a clean and strong interface [9, 10]. While aluminum alloys are the most common matrix employed in metal-ceramic composites, it is reported that the addition of TiC as a reinforcement improves the mechanical properties at both room and high temperatures. In particular, the Al-Cu alloy/TiC system provides a favorable combination of electrical and mechanical properties. Comparatively, Al-Mg alloy/TiC composites are lighter than Al-Cu alloy/TiC.

The interface between the matrix and reinforcement plays a crucial role in determining the properties of MMCs. Most of the mechanical and physical properties of the MMCs such as strength, stiffness, ductility, toughness, fatigue, creep, coefficient of thermal expansion, thermal conductivity, and corrosion are dependent on the interfacial behavior [11]. The main concern consists in the high corrosion tendency of Al alloys worsened by the galvanic corrosion between the metallic matrix and more noble fibers or particle reinforcement. Therefore, it is very important to study the corrosion behavior of composites for their applications.

The corrosion behavior of composites reinforced with different ceramics such as Al_2O_3, SiC, and TiC has been studied by Deuis et al. [12, 13] in 3.5 wt.% sodium chloride solution. They found that the corrosion rate increases in the following order: Al_2O_3 < SiC < TiC. Generally speaking, the corrosion rates of composites were higher than their matrix alloys when immersed in NaCl solutions. This has been attributed to localized attack of the MMC at the reinforcement matrix interface, resulting in pitting or crevice corrosion. The interfaces were preferred sites for passive film breakdown (pitting initiation sites) by promoting inhomogeneities that produce voids, resulting in an easier breakdown of the oxide layer [14, 15]. Pitting in composites has been observed at the reinforcement matrix interface [16, 17]. Factors influencing corrosion of the composites include porosity, precipitation of intermetallic phases within the matrix, high dislocation densities at the particle-matrix interfaces, the presence of an interfacial reaction product, and electrical conductivity of the reinforcements.

Sun et al. [18] studied the corrosion behavior of silicon carbide/aluminum (SiC_p/Al-6061) metal matrix composites in chloride solution by means of electrochemical techniques, scanning electron microscopy (SEM), Auger electron spectroscopy (AES), energy dispersive spectroscopy (EDS), and X-ray diffraction. Pit morphology was observed after anodic polarization to a number of potentials. It was seen that the corrosion potentials did not vary greatly or show definite trends in relation to the amounts of SiC_p reinforcement.

Although research has been carried out on the corrosion behavior of composites, some controversies still exist. The literature [19, 20] on corrosion behavior of metal matrix composites has shown that the presence of reinforcement may or may not increase the corrosion susceptibility, depending not only on the metal-reinforcement combination but also on the manufacturing process parameters. Trzaskoma et al. [19] studied SiC$_w$/Al-5456 and SiC$_w$/Al-6061 in order to compare pitting processes of SiC$_w$/Al metal matrix composites with that of corresponding unreinforced alloys. Studies of pit structure suggest there are two stages in pit development. The first involves the initial dissolution of metal atoms and opening of the pit, and the second involves pit enlargement or growth. For both materials, pits initiate at secondary particles within the metal matrix.

The use of Al-Cu alloys to fabricate MMCs by liquid metal infiltration can lead to the formation of the intermetallic CuAl$_2$ after the solidification and some other intermetallics, which not only increase some mechanical properties of the alloy but also can act as local anodes or cathodes, inducing, thus, a high susceptibility to localized types of corrosion like pitting, intergranular attack, etc. Localized pitting in the vicinity of copper aluminide (CuAl$_2$) precipitates within the 2000 series of aluminum alloys also was observed [21].

6.1.1 Experimental Conditions

In this work was investigated the aqueous corrosion behavior of aluminum metal matrix composites reinforced with TiC particles produced by the melt infiltration technique. The composite systems studied were Al-2024/TiC, Al-Cu$_x$/TiC, and Al-Mg$_x$/TiC.

The Al alloys/TiC composites were produced by pressureless melt infiltration of Al alloys into TiC preforms. The average size of the reinforcing TiC powder was 1.12 μm. These powders were uniaxially pressed into bars of approximately 6.5 × 1 × 1 cm in size. After, the green bars were partially sintered under argon at 1250, 1350, and 1450 °C for 90 min obtaining preforms with 44–48% of porosity.

Infiltration was carried out at 1100, 1150, and 1200 °C under argon atmosphere for the 2024 commercial aluminum alloy. The chemical composition of Al-2024 alloy was reported elsewhere [7, 22]. Meanwhile, the binary aluminum alloys were infiltrated at 900 and 1000 °C, and preforms used were sintered at 1250 °C both under an argon atmosphere. In addition, some specimens were heat treated at 530 °C during 150 min, water quenched, and subsequently artificially aged (AA) at 190 °C for 12 h under argon atmosphere and naturally aged (NA) at room temperature for 96 h, respectively.

Electrochemical experiments were performed using an ACM Instruments potentiostat controlled by a personal computer. The applied potential was ranged from ±1000 mV with respect to the free corrosion potential, E_{corr} at a scan rate of 1 mV s^{-1}. A saturated calomel electrode and graphite electrode were used as reference and auxiliary electrodes, respectively. Before starting the experiments, the E_{corr} value was measured for approximately 30 min to ensure stability. Corrosion rates were calculated in terms of the corrosion current density, I_{corr}, by using linear polarization resistance curves (LPR). This operation was done by polarizing the specimen from +10 to −10 mV with respect to E_{corr}, at a scan rate of 1 mV s^{-1} to get the polarization resistance R_p. The relationship between I_{corr} and R_p was calculated using the Stearn-Geary [23] equation:

$$I_{corr} = \frac{b_a b_c}{2.3(b_a + b_c)} \cdot \frac{1}{R_p},$$ (6.1)

where b_a and b_c are the anodic and cathodic Tafel slopes obtained from the polarization curves. During the experiment the E_{corr} values were measured using a digital voltmeter at intervals of 10 min. Thus, the variation of E_{corr} and the polarization resistance, R_p, with time, were recorded. All tests were performed at room temperature (25 °C ± 2 °C). A concentration of 3.5 (wt.%) of sodium chloride was used as electrolyte and prepared from analytical grade reagents. Samples of composites were machined and used like working electrodes (WE) with an exposure area of 1 cm^2.

6.1.2   Microstructure of Composites

Microstructure of composites before and after the experiments was determined by scanning electron microscopy (SEM) model Philips XL-30. Figure 6.1 shows typical micrographs of samples sintered at 1250 °C and infiltrated at 1000 °C where it can be seen in the matrix some CuAl$_2$ precipitates and titanium carbide particles. This figure shows the microstructure of the as-infiltrated and Al alloy/TiC composites. The dark phase is the aluminum matrix, and the white bright phases are the TiC grains. Intermetallic phases such as CuAl$_2$, Ti$_3$Cu, Al$_3$Ti, Ti$_3$AlC, and Ti$_3$Al were identified for specimens in the as-infiltrated condition and heat-treated composites [17]. The same phase was detected for specimens either artificially or naturally aged except the CuAl$_2$ phase, which was dissolved into the matrix when a solution heat treatment was applied. The slow cooling rate after infiltration lead the formation of phases such as the Ti$_3$AlC which is an intermediate compound between the TiAl$_3$ and TiC. X-ray tests indicated that Mg forms a solid solution with the aluminum lattice and no reaction phase was detected apart from aluminum and titanium carbide, since magnesium does not form stable carbides [2, 4, 5, 18]. The presence of these precipitates, especially the Cu containing, is detrimental from the corrosion point of view, since the Cu-Al galvanic coupling, being Cu the cathode, will produce sites for localized type of corrosion such as pitting corrosion, corroding the matrix surrounding the particle.

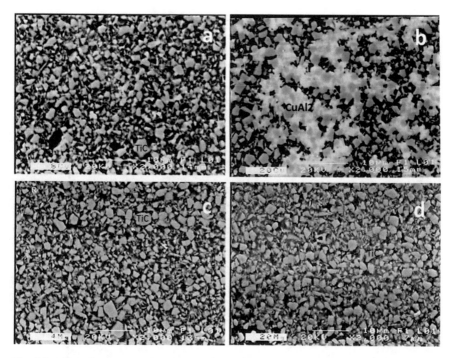

Fig. 6.1 Micrographs obtained by SEM of composites, (**a**) Al-1Cu/TiC, (**b**) Al-20Cu/TiC infiltrated at 1000 °C, (**c**) Al-1 Mg/TiC, (**d**) Al-20 Mg/TiC infiltrated at 900 °C

6.1.3 Electrochemical Evaluations

Figure 6.2a–c shows the effect of infiltration temperature on the anodic current density for the composites sintered at 1250, 1350, and 1450 °C, respectively. The densification of preforms increased with sintering temperature, which in turn decreases the anodic current density. For the specimens sintered at 1250 °C (Fig. 6.2a), the lowest anodic current density was obtained due to the lowest infiltrating temperature (1100 °C) and increased with increasing infiltration temperature. However, the anodic current density increased with higher sintered temperature (1450 °C) employed in specimens (Fig. 6.2c).

The specimens sintered at 1350 °C showed anodic current densities very similar or close for different infiltration temperatures as shown in Fig. 6.2b, but when the specimens were sintered at 1450 °C (Fig. 6.2c), the behavior was different, i.e., the highest anodic current density was observed at the lowest infiltrating temperature, it had the minimum value at the medium infiltrating temperature, and it had an intermediate value at the highest infiltrating temperature, just the opposite behavior observed with the specimens sintered at 1250 °C. The E_{corr} value remained almost unaffected in all these cases.

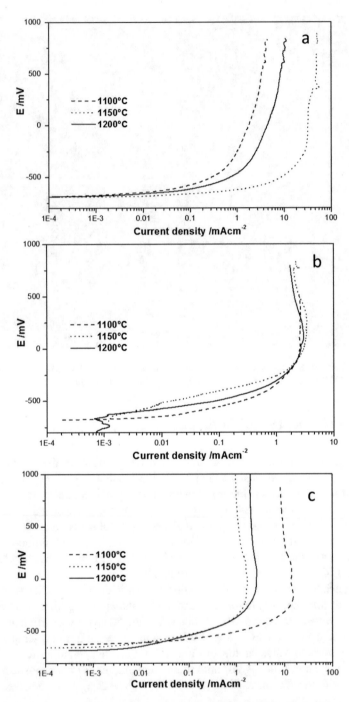

Fig. 6.2 Effect of the infiltration temperature on the anodic current density for Al-2024/TiC composites sintered at (**a**) 1250 °C, (**b**) 1350 °C, and (**c**) 1450 °C [1]

Table 6.1 Effect of heat treatment on the anodic current density for Al-2024/TiC composites

Composite	E_{corr} (mV)	I_{corr} (mA/cm^2)	CR (mm/year)
Al-2024/TiC as processed	−652	0.01	0.27
Al-2024/TiC artificially aged	−625	$3.1e^{-3}$	0.083
Al-2024/TiC naturally aged	−605	$1.5e^{-3}$	0.041

The effect of naturally or artificially aged composites on the polarization, current density, and corrosion rate is shown in Table 6.1, where it can be seen clearly that these treatments had a slight negative effect on the anodic current density on the polarization curve of the Al-2024 alloy but the passive region is kept. A slight increase, about two times, on the anodic current density is observed with these heat treatments. Although the CuAl$_2$ particles disappear by dissolution in the aluminum with these treatments [24], there are still some other intermetallics such as Ti$_3$Cu, Al$_3$Ti, Ti$_3$AlC, and Ti$_3$Al which are cathodic with respect to the metallic matrix.

The effect of Cu and Mg content in the binary alloys for polarization curves is shown in Fig. 6.3. Figure 6.3a shows the effect of Al-Cu$_x$ alloy in the composite on the polarization curves. This figure shows that for over-potentials less than 250 mV the anodic current density decreases with additions of Cu and the E_{corr} sometimes decreased but some other times increased; the opposite occurs when over potentials higher than 250 mV are applied, additions of Cu always increased the anodic current density. Figure 6.3b shows the effect of Al-Mg$_x$ alloy in the composite on the polarization curves. Similar behavior is observed in the cases of Al-Mg$_x$/TiC composites. In all cases, increasing the Mg content in composite increased the anodic current density values and decreased the E_{corr} values.

Preforms sintered at 1250 °C were infiltrated at three different temperatures (1100, 1150, and 1200 °C), and the I_{corr} was obtained in function of the time as is shown in Fig. 6.4. The I_{corr} values tend to decrease with time, reaching stable values after 16 h, except with the specimen infiltrated at 1100 °C after 15 h of exposure, which had a higher I_{corr} value.

In all cases the corrosion current density changes were recorded during a 24-h period. The corrosion current density variations can be observed in function of time as is shown in Fig. 6.5a, b. In Fig. 6.5a, for Al-8Cu/TiC during a 20-h period, the corrosion current density decreased lightly showing a change in I_{corr} value probably due to a corrosion film product breakdown which brings a corrosion current increment. However, it was observed a diminishing tendency to decrease in time (forming a new film again), meanwhile other alloys keep a constant I_{corr} value. Since the effect of different composite condition in I_{corr} was the same as the one observed in the polarization curves, here it shows only some examples. As an example, in Fig. 6.5a, b, it is observed the effect of additions of Cu and Mg compared to the Al-2024 alloy. As in the polarization curves, the addition of these elements increased the I_{corr} values up to almost ten times. In most of the cases, the I_{corr} values remained quite stable during exposure time, which may be due to a protective Al$_2$O$_3$ or Al(OH)$_3$ layer.

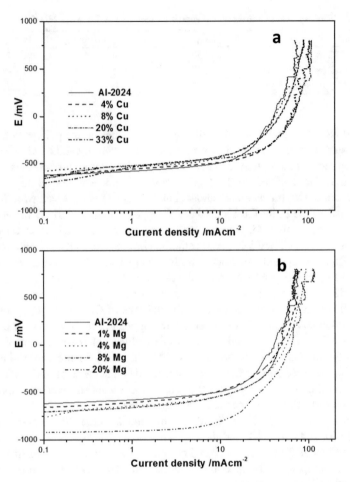

Fig. 6.3 Effect of (**a**) Cu and (**b**) Mg additions on anodic dissolution of the Al-Cu$_x$/TiC and Al-Mg$_x$/TiC composites [1]

6.1.4   Surface Analysis by SEM

Figure 6.6 shows the surface morphology of the Al-2024/TiC composite after immersion in 3.5 wt.% NaCl solution; a large number of pits are evident. The different sintering and infiltrating temperatures decreased the number of pits, but they did not disappear completely, as shown in Fig. 6.7, where localized corrosion is observed.

Figures 6.8 and 6.9 show the effect of Cu or Mg additions in composites which produce hollows and big diameter pits with a bigger associated with the localized attack of a protective Al$_2$O$_3$ or Al(OH)$_3$ layers by chloride ions [15–17]. The corrosion rate is accelerated by other factors such as the galvanic coupling of Cu$^+$- or Ti$^+$-containing particles in Al matrix.

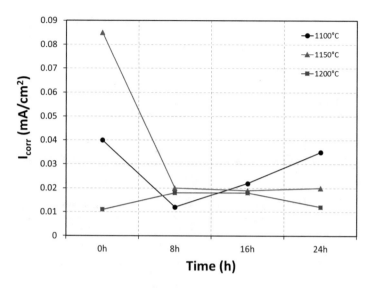

Fig. 6.4 Effect of the infiltration temperature on I_{corr} as a function of the time for Al-2024/TiC composites

The nucleation of these pitting is in the TiC matrix and intermetallic compound-matrix interfaces. These sites act as discontinuities during the establishment of the protective film or passivation, and can be easily disrupted in that particular site. A quick dissolution is carried out inside the pitting while there is an existing oxygen reduction process over adjacent surfaces. There is a high concentration of Al^{3+} ions inside the pitting as a result of cationic hydrolysis, and there is an increment of hydrogen ions according to the following reaction:

$$Al^{3+} + 3H_2O \rightarrow Al(OH)_3 + 3H^+ \qquad (6.2)$$

This process reduces the pH inside the pit, and material transfer to the adjacent zone occurs by ion transport in the electrolyte. Intermetallic Cu–Al compounds are strongly cathodic inside the matrix and act as cathodic sites, facilitating the disruption of either Al$_2$O$_3$ or Al(OH)$_3$ protective layers and enhancing pitting corrosion. When second phases are present, such as Cu- or Ti-rich particles, they act as effective cathodes, leading to the formation of micro-galvanic cells, enhancing the anodic dissolution of the metallic matrix in the form of pits.

In the case of the Al-2024/TiC composites, the anodic current density was reduced, decreasing the amount and size of the pits. This was independent of the sintering and infiltrating temperatures. For composites, either artificially or naturally aged, the anodic corrosion current density increases, and the number and depth of pits decrease. It was observed that there is no direct relationship between the anodic current density and the sintering or infiltrating temperatures.

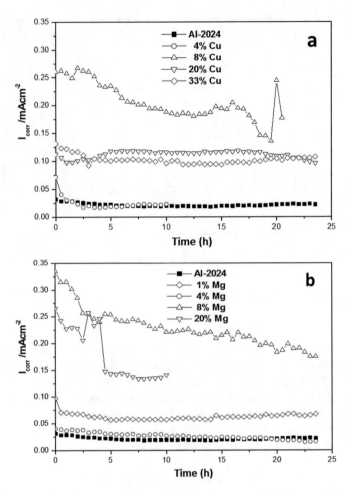

Fig. 6.5 Effect of additions of (**a**) Cu and (**b**) Mg on I_{corr} vs time obtained of the Al-Cu$_x$/TiC and Al-Mg$_x$/TiC composites [1]

The galvanic effect of Cu- and Ti-rich particles to the Al alloy enhances pitting corrosion. Meanwhile, the anodic current density in the composites increases with the addition of either Cu or Mg.

6.2 Corrosion Behavior of Mg-AZ91E/TiC Composites

The first composite studies were based in aluminum matrix, but recently magnesium and its alloys have been used extensively in the fabrication of composites. Research in new systems is required due to rapid increase in technological development. Therefore, it could be appreciated as an increasing interest in the use of magnesium

Fig. 6.6 Micrograph
obtained by SEM of
Al-2024/TiC composite
corroded in 3.5% NaCl,
where some pits can be
observed

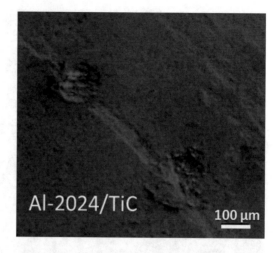

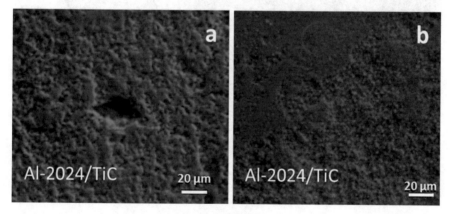

Fig. 6.7 Micrograph obtained by SEM of Al-2024/TiC composite sintered at 1450 °C and
infiltrated at (**a**) 1100 °C and (**b**) 1200 °C corroded in 3.5% NaCl

and its alloys as a metallic matrix for MMC composites. Nowadays, aluminum and
magnesium matrixes are widely used in metallic matrix composites (MMCs),
because they have the highest priority in applications where a combination of
corrosion resistance, low density, and high mechanical performance is required,
such as in the automotive and aerospace industry.

The reinforcement of magnesium matrix, based on the use of TiC particles, is
interesting because as reinforcement it improves the mechanical properties at room
and high temperatures [7–10]. The main disadvantage of magnesium is the high
chemical reactivity due to its negative electrochemical potential; this greatly restricts
its industrial applications, and the same disadvantage has been found for Mg-Al-Zn
alloys, being Mg-AZ91E, one of the most used alloy.

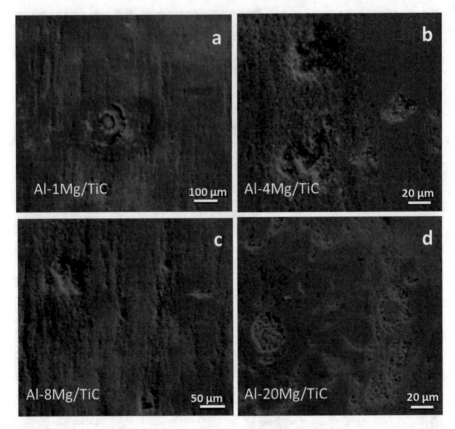

Fig. 6.8 Micrographs obtained by SEM of Al-Mg/TiC composites corroded in 3.5% NaCl solution; (**a**) Al-1 Mg/TiC, (**b**) Al-4 Mg/TiC, (**c**) Al-8 Mg/TiC, (**d**) Al-20 Mg/TiC

Currently, there are numerous research on MMC using magnesium and its alloys like matrix [9, 10, 25–32]. Many studies have been carried out to determine the corrosion behavior.

Pardo et al. [33] concluded that corrosion damage was mainly caused by formation of $Mg(OH)_2$ corrosion film. Mg-AZ80 and Mg-AZ91 are magnesium alloys resistant to corrosion. The relative fine-phase ($Mg_{17}Al_{12}$) network and the aluminum enrichment produced on the corroded surface were the key factors limiting progression of the corrosion attack. Preferential attack was located at the matrix β-phase and MnAl intermetallic compounds. Nuñez et al. [34] reported the corrosion behavior of Mg-ZC71/SiC composites with 12 vol. % SiC, by electrochemical tests. They found that MMC composite showed ten times difference in corrosion rates than the alloy. Local corrosion was three times faster in the composite than in the monolithic alloy. Suqiu et al. [35] reported the corrosion resistance in Mg-AZ91D/TiC composites. They observed that the local corrosion rate was higher in the composite due the formation of a thinner layer of corrosion products than in the monolithic alloy.

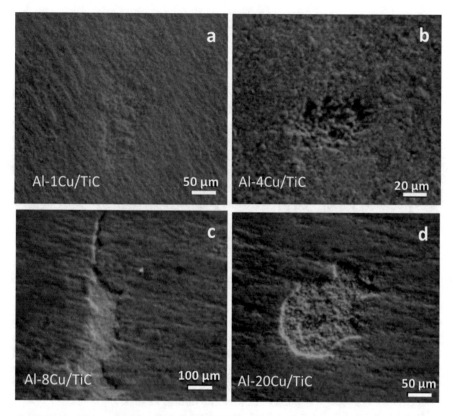

Fig. 6.9 Micrographs obtained by SEM of Al-Cu/TiC composite corroded in 3.5% NaCl solution; (**a**) Al-1Cu/TiC, (**b**) Al-4Cu/TiC, (**c**) Al-8Cu/TiC, (**d**) Al-20Cu/TiC

Tiwari et al. [36] studied the corrosion behavior of two SiC Mg-based MMCs, namely, Mg-6SiC and Mg-16SiC (vol. %), in aerated 1MNaCl and compared them with the corrosion rate of pure Mg. The presence of SiC particles deteriorated the corrosion resistance of magnesium, and the corrosion resistance decreased with increasing SiC volume fraction, finding that galvanic corrosion between Mg matrix and SiC reinforcement did not significantly contribute to the overall corrosion rate.

Salman et al. [37] carried out a comparative electrochemical study of AZ31 and AZ91 magnesium alloys in 1 M NaOH and 3.5 wt. % NaCl solutions at room temperature. In 1 M NaOH solution, AZ31 alloy showed several potential drops throughout the experiment, but AZ91 alloy did not show. When anodized at 3 V for 30 min in 1 M NaOH solution, the anticorrosion behavior of anodized specimens was better than those of specimens which were not anodized. In similar way Singh et al. [38] study the corrosion behavior of Mg, AZ31, and AZ91 in 3.5% NaCl solution using weight loss, electrochemical polarization, and impedance measurements. The corrosion current density (I_{corr}) derived from the Tafel plots exhibited their corrosion resistances in the order of Mg > AZ91 > AZ31.

Table 6.2 Chemical composition of the Mg-AZ91E alloy (wt.%)

Al	Mn	Zn	Si	Fe	Cu	Ni	Mg
8.7	0.24	0.7	0.2	0.005	0.015	0.001	Bal.

Budruk et al. [39] studied the corrosion behavior of pure magnesium, Mg-Cu (0.3, 0.6, and 1 vol. %), and Mg-Mo (0.1, 0.3, and 0.6 vol. %) composites in 3.5% NaCl solution, finding that the corrosion rate increased with increasing the volume fraction of reinforcement in both composites. Microscopic observations of corroded specimens confirmed micro-galvanic activity at the matrix-reinforcement interfaces.

In this research work, corrosion behavior of Mg-AZ91E alloy and Mg-AZ91E/ TiC composites in NaCl solution was carried out. It was studied the effect of TiC-reinforcing particles on the corrosion resistance of Mg-AZ91E alloy in 3.5% NaCl solution [32].

6.2.1   *Experimental Conditions*

Chemical composition of Mg-AZ91E alloy is shown in Table 6.2. In order to fabricate the composites, first we sintered preforms, using TiC powders with an average particle size of 1.1 2 μm. These powders were sintered in an argon atmosphere at 1250 °C during 60 min, obtaining densities of 56% vol. Composites were fabricated by pressureless melt infiltration of Mg-AZ91E alloy into TiC porous preform (44%) at 950 °C in argon atmosphere during 12 min.

For the corrosion tests, a 3.5% NaCl solution was used at room temperature (25 °C). Polarization curves were recorded at a constant sweep rate of 1 mV s^{-1} from −800 to +800 mV interval with respect to open circuit potential (E_{corr}). A conventional three-electrode glass cell was used with a graphite rod as auxiliary electrode and a saturated calomel electrode (SCE) as reference.

Corrosion current density values (I_{corr}) were calculated by using the Tafel extrapolation method and by taking an extrapolation interval of ±250 mV around the E_{corr}. Linear polarization resistance (LPR) measurements were carried out by polarizing the specimen from +10 to −10 mV in respect to E_{corr}, at a scanning rate of 1 mV s^{-1} every 60 min during 24 h. Electrochemical impedance spectroscopy tests were carried out at E_{corr} by using a signal with an amplitude of 10 mV and a frequency interval of 0.1 Hz to 10 kHz. An ACM potentiostat controlled by a desktop computer was used for the LPR tests and polarization curves, whereas for the electrochemical impedance spectroscopy (EIS) measurements, a model PC4 300 Gamry potentiostat was used.

Electrochemical noise (EN) measurements in both current and potential were recorded using two identical working electrodes and a reference electrode (SCE). The electrochemical noise measurements were made recording simultaneously the potential and current fluctuations at a sampling rate of 1 point per second during a period of 1024 s. Removal of the direct current trend from the raw noise data was the first step in the noise analysis when needed. To accomplish this, a least square fitting method was used. Finally, the noise resistance (Rn) was then calculated as the ratio of the potential noise standard deviation (σ_v) over the current noise standard deviation (σ_i), according to:

$$Rn = \sigma_v/\sigma_i, \tag{6.3}$$

where Rn can be taken as the linear polarization resistance, Rp in the Stern-Geary equation (6.1).

6.2.2   Microstructure of Composites

Microstructures of Mg-AZ91E alloy and Mg-AZ91E/TiC composite are shown in Fig. 6.10. Figure 6.10a shows the surface image obtained by scanning electron microscopy (SEM) showing the primary α-Mg phase, together with an eutectic phase aluminum rich surrounding the β phase constituted of $Al_{12}Mg_{17}$. Additionally, some AlMn-base precipitates could be observed. Figure 6.10b shows the microstructure of AZ91E/TiC composite.

6.2.3   Electrochemical Evaluations

Polarization curves for Mg-AZ91E alloy and Mg-AZ91E/TiC composite exposed to 3.5% NaCl solution is shown in Fig. 6.11. These curves showed an active behavior, without any passive zone, instead, an anodic limiting current was observed, with values of 6.11 and 0.16 A/cm^2 for AZ91E alloy and AZ91E/TiC composite,

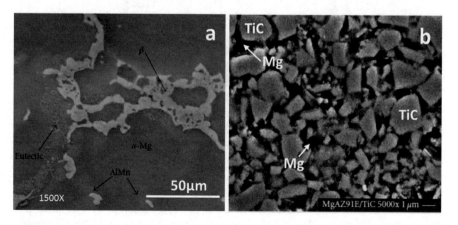

Fig. 6.10 SEM images of the microstructure of (**a**) Mg-AZ91E alloy and (**b**) Mg-AZ91E/TiC composite [32]

Fig. 6.11 Polarization
curves for Mg-AZ91E alloy
and AZ91/TiC composite
exposed to 3.5% NaCl
solution [32]

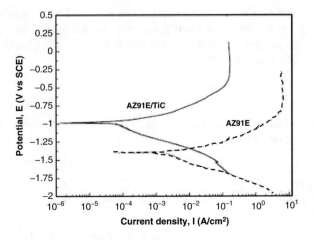

Fig. 6.11 Polarization curves for Mg-AZ91E alloy and AZ91/TiC composite exposed to 3.5% NaCl solution [32]

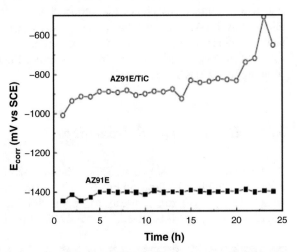

Fig. 6.12 E_{corr} versus time for AZ91E alloy and AZ91E/TiC composite exposed to 3.5% NaCl solution [32]

respectively. The E_{corr} value was more active for the base alloy, close to -1390 mV than the composite which has a value of -980 mV.

The corrosion current density value was also lower for the composite, 8.8×10^{-5} A/cm^2, than that for the base alloy, 2.9×10^{-3} A/cm^2. Figure 6.12 shows the change in the E_{corr} value with time; Mg alloy kept more active values than those for the composite, although the later had a tendency toward noble values.

The change in the Rp value with time for both the base metal and composite is shown in Fig. 6.13, where it can be observed that AZ91E alloy showed lower Rp values and, thus, higher corrosion rates than the AZ91E/TiC composite. However, the Rp value for the AZ91E base alloy showed a tendency to increase and after that to decrease as time elapsed, and thus, the corrosion rate decreases first and after that it increases, indicating the cracking or detachment of any protective corrosion product layer, showing the non-protective nature of this film. It could be also due

Fig. 6.13 Polarization resistance (Rp) versus time for AZ91E alloy and AZ91E/TiC composite exposed to 3.5% NaCl solution [32]

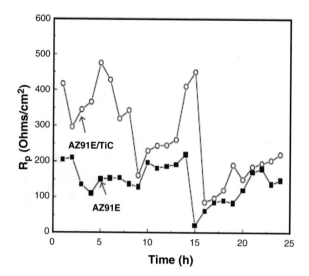

to the fact that the $Al_{12}Mg_{17}$ particles enhance the dissolution of the α-Mg phase since it is cathode as compared to the matrix.

Figure 6.14a shows the Nyquist diagrams for the AZ91E alloy, which can be observed as a capacitive semicircle at high frequencies, but at lower or intermediate frequencies, the data described a straight line, indicating that the corrosion process is under a mixed mechanism: charge transfer from the metal to the interface through the double electrochemical layer and by the diffusion of aggressive ions through the corrosion products layer. The diameter of the high frequency capacitive semicircle, described as the charge transfer resistance (R_{ct}) equivalent to the polarization resistance (Rp), decreased as time elapsed, with an increase in the corrosion rate with time.

Figure 6.14b shows the Nyquist diagrams for AZ91E/TiC composite. This figure showed a capacitive-like, depressed semicircle at high frequencies, followed by an inductive loop at low and intermediate frequencies, indicating that the corrosion process is under control of adsorption of chloride ions at the metal/solution interface. The diameter of the semicircles decreased as time elapsed, indicating an increase in the corrosion rate with time and the non-protective nature of the corrosion products.

There is a factor called localization index (LI) defined as:

$$LI = \sigma_i / IRMS, \tag{6.4}$$

where σ_i is the current noise standard deviation and IRMS is the current root mean square value [40–42].

LI is the distribution measure data around of the root mean square of current (IRMS). It has values from 0 to 1. Table 6.3 shows the value range for LI and its relation with the corrosion type according to the morphology of the attack.

Fig. 6.14 Nyquist diagrams in function of time for (**a**) Mg-AZ91E alloy and (**b**) AZ91E/TiC composite exposed to 3.5% NaCl solution [32]

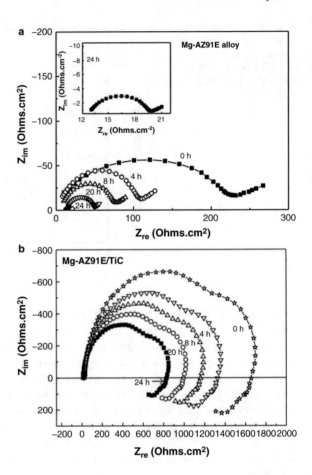

Table 6.3 Correlation between localized index (LI) and corrosion type

LI values range	Corrosion type
0.001–0.01	General corrosion
0.01–0.1	Mix corrosion
0.1–1.0	Localized corrosion

LI values for AZ91E base alloy and AZ91E/TiC composite were 0.78 and 0.2, respectively. According to this LI values, both Mg alloy and composite have a localized corrosion type.

By using the potential noise standard deviation (σ_v) and dividing it over the current noise standard deviation (σ_i), we can obtain a noise resistance (Rn). Figure 6.15 shows the noise resistance versus time for AZ91 E alloy and composite. This figure shows that Rn decreased as time elapsed. The AZ91E alloy exhibited lower Rn values than those exhibited by the composite. Thus, AZ91E alloy has higher corrosion rate than composite.

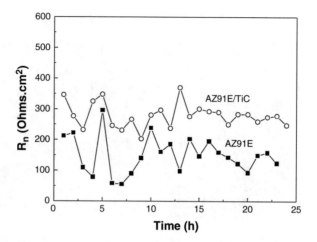

Fig. 6.15 Noise resistance in function of time for Mg-AZ91E alloy and AZ91E/TiC composite exposed to 3.5% NaCl solution [32]

Once hydrolysis reaction starts, corrosion reaction proceeds by the oxidation of Mg as Mg^{2+} and their involvement with OH^- ions that is produced by the reduction of water molecule [38, 43]. This results in the formation of $Mg(OH)_2$ precipitates at the metal surfaces. The reaction involves in the formation of $Mg(OH)_2$ according to the following reactions:

$$Mg^{2+} + 2OH^- \rightarrow Mg(OH)_2 \tag{6.5}$$

$$Mg^{2+} + 2H_2O + O_2 \rightarrow 2Mg(OH)_2 \tag{6.6}$$

Similar reactions may also be involved in the corrosion of Al and Zn of the alloys. Formation of $Mg(OH)_2$ with the increase of potential in the anodic region and their subsequent accumulation at the electrode interfaces seems to be the main reason for the occurrence of the diffusion-limiting behavior in the anodic region of the alloy and composite.

Scanning electron microscopy (SEM) images of Mg-AZ91E alloy and Mg-AZ91E/TiC composite of corroded specimens are shown in Fig. 6.16. Preferential attack along the matrix and the TiC particle interfaces can be seen due to a different electrochemical potential between the α-Mg, β-Mg, AlMn inclusions, and TiC particles, which gives a galvanic effect.

Corroded surface shows the accumulation of corrosion product over the complete surface and mainly in the grain boundaries. Since grain boundaries are thermodynamically more active, precipitation of corrosion product $Mg(OH)_2$ in this region reduces the corrosion attack. This is perhaps reason for owing better corrosion resistances by Mg metal. Some researchers have reported that corrosion resistance of composites is higher than for their matrix alloy [44].

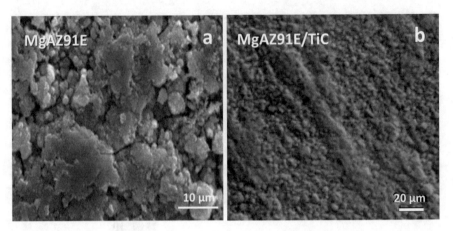

Fig. 6.16 SEM micrograph of corroded (**a**) Mg-AZ91E alloy and (**b**) Mg-AZ91E/TiC composite

6.3   Corrosion Behavior of Ni/TiC Composites

This work shows the electrochemical study of nickel (Ni) and Ni/TiC composite immersed in synthetic seawater [45, 46]. In order to characterize the corrosion process, polarization curves (PC) were carried out. All electrochemical measures were made under static conditions, room temperature, atmospheric pressure, and 24-h exposure time. An electrochemical cell with a typical three-electrode array was used. Effect of TiC as reinforcement into the Ni matrix was evaluated.

Although there are many metal matrix composite (MMC) fabrication methods like powder metallurgy, squeeze casting, stir casting, thermal spray, and vortex method among others, the infiltration process is one of the most widely used method [1, 9, 10, 22, 30–32]. The infiltration process of a ceramic preform without external pressure has some advantages for the fabrication of composites such as isotropic properties, which in turn will be reflected in the mechanical properties and corrosion resistance.

During the infiltration process, the wettability of metal (in this case Ni) on the reinforcement (TiC) plays an important role because this phenomenon defines the strength of the matrix-reinforcement interface [7, 8, 11]. Poor wetting creates sites with weak interfaces that can be sites for initiation of crevice corrosion.

Aluminum, magnesium, copper, and nickel are the metals most used like matrix in the composites. There are many combinations of reinforcements and matrix, such as Al/TiC, Mg/TiC, Mg/SiC, Al/SiC, Al/Al$_2$O$_3$, and Al/AlN, among others [1, 9, 11, 22, 24–31, 47–50]. However, one of the ceramics most used currently like reinforcement is TiC, in most of the cases combined with Al and Mg alloys and more recently with Cu and Ni.

Ni/TiC composite was obtained to satisfy some applications where the materials need high mechanical properties and corrosion resistance. In order to study the electrochemical behavior of the Ni/TiC, seawater as electrolyte was used. This seawater is a high corrosive environment, and the corrosion of the metallic structures in this electrolyte is influenced by many factors, such as temperature, pH, salinity, etc. [51]. It is important to point out that the electrochemical technique of polarization curves (PC) was used in the corrosion study of this composite. In addition, the corrosion potential values with respect to the time were measured.

6.3.1 Experimental Conditions

A composite material containing 60% of TiC like reinforcement (99.9% H.C. Starck, $D_{50} = 4.23$ μm) and 40% of Ni like a matrix was used in this study. The composite material was processed by pressing TiC powders in a metallic die in order to obtain green preforms in the form of small bars with dimensions of 6x1x1 cm, after these preforms were sintered at 1250 °C for 1 h in a tubular furnace with a flux of argon to avoid oxidation. Later, the preforms sintered were infiltrated with Ni at temperature of 1515 °C during 15 min under a flux of argon.

Ni/TiC composites and Ni were machined in order to be used like working electrode (WE) in all the tests. The total exposure area of this WE was 1 cm². Before each test, the WE was polished with silicon carbide (SiC) paper up to 600 grit and after was cleaned with deionized water and degreased with acetone.

The test electrolyte used was a synthetic seawater solution, which was prepared in accordance with ASTM D-1141 [52]. A three-electrode electrochemical cell was used: working, reference, and auxiliary electrode. Working electrode (WE) was made from nickel and composite samples, the reference electrode was a saturated calomel electrode (SCE), and a sintered graphite rod was used as auxiliary electrode (AE). All electrochemical measurements were carried out at atmospheric pressure and room temperature of Veracruz Port, México. The electrochemical study was performed with an ACM Instruments Gill AC potentiostat/galvanostat, which was coupled to a computer for each of the electrochemical tests.

Polarization curves according to ASTM G5 [53] were obtained. To obtain the PC for both Ni/TiC composite and Ni, polarizations of ± 0.5 V with respect to the corrosion potential (E_{corr}) were carried out, with a sweep speed of 0.001 V/s. The measurements were carried out under static conditions, atmospheric pressure, and room temperature.

6.3.2 Composite Microstructure

Figure 6.17 shows the microstructure obtained by SEM of Ni/TiC composite used in this study. The structure consists of small TiC particles immersed in a Ni matrix. A homogeneous distribution of TiC particles can be observed.

Fig. 6.17 Typical
microstructure of Ni/TiC
composite

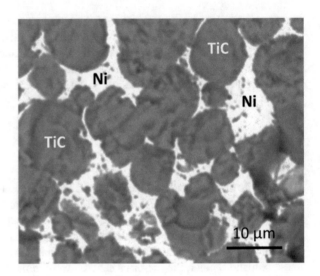

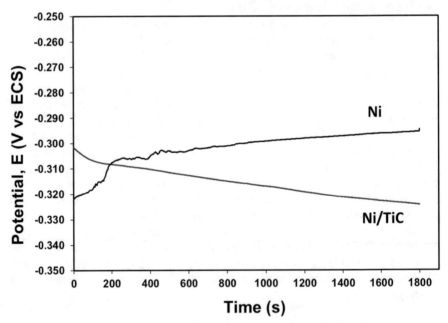

Fig. 6.18 Open corrosion potential (OCP) for nickel and Ni/TiC composite [45, 46]

6.3.3 Electrochemical Evaluations

Prior to the completion of electrochemical techniques, Ni/TiC composite and Ni
specimens were corroded freely, performing potential measurements until stability
was achieved. The E_{corr} measurements are shown in Fig. 6.18.

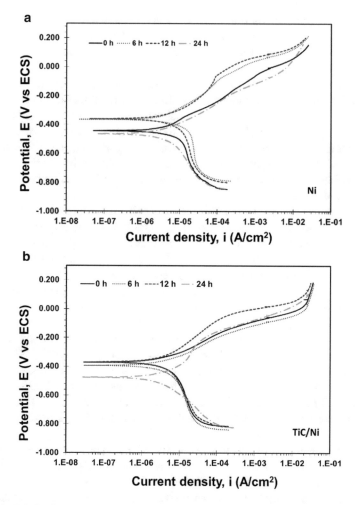

Fig. 6.19 Polarization curves for (**a**) Ni and (**b**) Ni/TiC composite at different immersion times [45, 46]

The profiles of E_{corr} tests shown in Fig. 6.18 show that the stabilization period for both materials is 30 min (1800 s). This stabilization must be achieved prior to the completion of any electrochemical technique. Also it can be observed that the E_{corr} of the Ni oscillates in values close to -0.290 V vs ESC, whereas the E_{corr} of Ni/TiC composite material is about -0.320 V vs ECS, approximately. This variation in the potential values between Ni/TiC and Ni shows that the E_{corr} of Ni is directly influenced by the addition of the TiC ceramic.

Figure 6.19 shows the polarization curves obtained for Ni/TiC composite and Ni immersed in synthetic seawater at different immersion times. In the PC for Ni/TiC, it is possible to appreciate that the E_{corr} is maintained in values of near potentials; however when reaching the 24 h of immersion, a displacement to values of more electronegative potentials appears. These effects are explained in terms of the ability of the chloride to infer in the formation and repair of the own passive film of the Ni matrix of the composite and thus to cause a displacement of the E_{corr} to more cathodic values [54, 55].

In the anodic branch of PC for Ni/TiC, there is a tendency of passivation from 0 to 6 h reaching small areas of passivation at 12 and 24 h. However, the formation of corrosion products beneficial for the material to be passive is not very marked, so it is possible to say that in the oxidation reaction of the metal there is a process of charge transfer (activation process) influenced by a process of mass transfer (diffusional process). This behavior is attributed to the structural characteristic of the composite, presenting interstitial sites, capable of generating differential aeration cells that cause active sites of corrosion.

In the cathodic branch of the PC of the composite, it is observed in a similar way that in all the times of exposition, a load transfer process with mass transfer influence is presented. The corrosion process that is presented is of a mixed type.

In the PC for Ni, it is observed how the E_{corr} moves to values of less electronegative potentials from 0 to 6 h; later the potential is shifted to more electropositive values as the exposure time elapses. This beahavior is attributed to the Cl-containing medium involved in the mechanism of generation and dissolution of the film of corrosion products [55, 56].

In the anodic branch of the PC for Ni, a similar behavior occurs in all the immersion times, reaching small areas of passivation; however the Ni does not passivate completely; this coupled with the slopes of Tafel that oscillate in approximately 0.180 V, indicating that load transfer processes with mass transfer influence are presented. For the cathodic branch, it is observed that at 6 and 24 h, there is a process of mass transfer, while at 0 and 12 h of immersion, a mixed process is presented.

The electrochemical parameters of PC for Ni/TiC composite were determined from the Tafel extrapolation method, resulting in the values presented in Table 6.4. As can be observed, after 6 h of immersion, the Tafel slopes recorded values higher than 0.120 V, corroborating that the corrosive process present in the composite is of mixed type, as observed in Fig. 6.19.

The electrochemical parameters of PC for Ni were determined from the Tafel extrapolation method, resulting in the values presented in Table 6.5. As can be observed for all times of immersion, with the exception of 6 and 24 h where a purely active process occurs, the slope values are higher than 0.120 V, that is, there are influenced transfer processes by mass transfer.

Table 6.4 Parameters obtained from PC of Ni/TiC composite exposed to seawater

Time (h)	ba (V/decade)	bc (V/decade)	β (V)	I_{corr} (A/cm^2)	CR (mm/year)
0	0.110	0.132	0.0260	6E−06	0.0646
6	0.167	0.361	0.0495	5E−06	0.0538
12	0.192	0.253	0.0473	7E−06	0.0753
24	0.176	0.158	0.0361	7E−06	0.0753

Table 6.5 Parameters obtained from PC of Ni exposed to seawater

Time (h)	ba (V/decade)	bc (V/decade)	B (V)	I_{corr} (A/cm^2)	CR (mm/year)
0	0.163	0.157	0.0347	6.1E−06	0.0657
6	0.179	0.105	0.0287	5.0E−06	0.0538
12	0.153	0.148	0.0326	8.0E−06	0.0861
24	0.124	0.114	0.0257	2.5E−06	0.0269

Fig. 6.20 Corrosion rate for Ni/TiC composite and Ni at different immersion times [45]

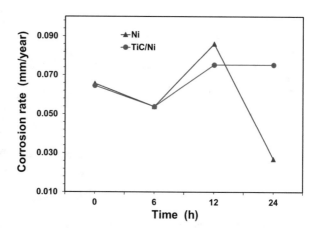

The above corroborates that the corrosive process present in the Ni is of the mixed type.

The values of corrosion rate (CR) for the Ni/TiC composite and Ni during the first 24 h of immersion were calculated from the I_{corr} obtained from the PC presented in Tables 6.4 and 6.5 and are shown in Fig. 6.20.

In Fig. 6.20 it can be observed that at the beginning of the immersion time from 0 to 6 h, the CR behavior is similar for Ni/TiC and Ni, which can be attributed in both cases to the rupture of the film of the corrosion products, due to the Cl⁻ [54, 57] ions adsorbed on the metal surface causing the increase of CR, to later regenerate for the case of Ni and decrease the values of CR continuously until reaching 24 h, which does not happen in the composite. This behavior can be attributed to the porosities in the composite, which cause localized sites where the active area increases.

The localization index (LI) defined in Eq. (6.4) was used to calculate the type of corrosion for Ni and Ni/TiC composites. LI is the distribution measure data around of the root mean square of current (IRMS). It has values from 0 to 1. Figure 6.21 shows the value range for LI and its relation with the corrosion type according to the morphology of the attack for Ni and Ni/TiC composites.

In the case of Ni, a localized corrosion type was observed, except for 24 h where a mix corrosion type belongs. For the Ni/TiC composites, a general corrosion type was obtained according to LI.

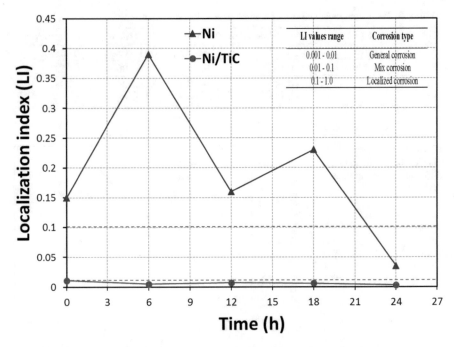

Fig. 6.21 Corrosion rate for Ni/TiC composite and Ni at different immersion times

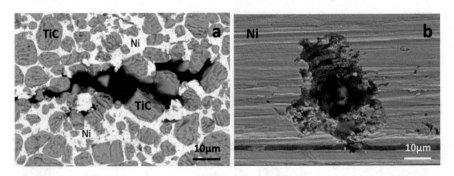

Fig. 6.22 Images obtained by SEM of electrodes surface corrode surface after been exposed 1 day in seawater, (**a**) Ni/TiC composite, (**b**) Nickel [46]

6.3.4   Surface Analysis by SEM

Surface analysis of Ni/TiC composite and Ni matrix samples were exposed in synthetic seawater for immersion periods of 1 day. Subsequently, each of the samples was cleaned according to ASTM D G1-2011 [58] to be characterized by SEM in order to know its morphology. Figure 6.22 shows images obtained by SEM of the surface electrodes of Ni/TiC composite and Ni.

The surface analysis of the morphology of the corrosion process indicates that the corrosive attack carried out on the Ni/TiC composites was mix in the first hours of exposure, and as the time evolve, a localized (crevice type) corrosion was observed, which could be attributed to the pores of the composite material or sites where the Ni does not infiltrate completely, which cause interstitial sites between Ni and TiC, which in turn are preferential sites to initiate the corrosion process. The corrosion mechanism of the composite is a differential aeration cell (Fig. 6.22a).

For the case of Ni electrodes after being exposed to seawater for 1 day, in most of cases, it exhibited a localized corrosion type, specifically pits (Fig. 6.22b). And in some cases, a mix corrosion type was observed.

According to CR results of Ni and Ni/TiC composites immersed in synthetic seawater, the values of CR corresponding to Ni/TiC composite exposed for long periods of time are higher than the CR of the Ni used as matrix. This behavior could be attributed to the crevice corrosion caused by TiC particles and microvoids in the network.

6.4   Corrosion Behavior of Al-Cu/TiC Composites

This work shows the electrochemical study of Al-Cu/TiC composite and Al-Cu alloy immersed in synthetic seawater. Polarization curves (PCs) and the corrosion potential as a function of the time were the electrochemical techniques used to characterize the corrosion process [59].

The composite materials have been displaced to the commercial steels, mainly when specific properties are needed such as better resistance to corrosion, high toughness, and wear resistance among others. In previous works Albiter et al. [1] studied the corrosion resistance of composites in 3.5% NaCl solution evaluated as a function of the addition of Cu and Mg into the aluminum. In all cases pitting corrosion was observed with or without additions of Cu or Mg, but these elements increased the anodic corrosion current. In similar way, the corrosion behavior of composites reinforced with different ceramics, such as Al_2O_3, SiC, and TiC, has been studied by Deuis et al. [12] in 3.5 wt.% sodium chloride solution. They found that the corrosion rate increases in the following order: Al_2O_3 < SiC < TiC. The corrosion rates of composites were higher than their matrix alloys when they were immersed in NaCl solutions.

Al-Cu/TiC composite was obtained to satisfy specific mechanical properties. In order to generate information about the corrosion kinetic of this composite, synthetic seawater was used as corrosive environment, and using electrochemical tests, the corrosion behavior was evaluated.

6.4.1 Experimental Conditions

Porous preforms were prepared by uniaxially pressing TiC powders (H.C. Starck grade c.a.s.) with an average size particle of 4.2 µm, in a rectangular die to form green bars of 6.5 × 1 × 1 cm in size. These preforms were sintered at 1250 °C for 1 h in a tube furnace under flowing argon, obtaining porous preforms with 60% of theoretical density. The aluminum alloy (Al-4%Cu) was used to infiltrate the porous preforms at 1100 °C in an argon atmosphere. The final Al-Cu/TiC composites have a volumetric fraction of about 60% of TiC and 40% of Al-Cu alloy.

Samples of composite and Al-Cu alloy were machined and used like working electrodes (WE) with an exposure area of 1 cm^2. Synthetic seawater was used as test environment. This solution was made according to ASTM D1141 [52]. The exposure time in all corrosion tests was 24 h at room temperature, atmospheric pressure (Veracruz Port, Mexico), and static conditions.

A three-electrode (working, reference, and auxiliary electrode) electrochemical glass cell was used. A saturated calomel electrode (SCE) and a sintered graphite rod were used as reference (RE) and auxiliary electrode (AE), respectively. In order to minimize the effect of the solution resistance, a Lugging capillary was used.

The electrochemical techniques used were corrosion potential (E_{corr}) and potentiodynamic polarization curves (PCs). PCs were recorded at a sweep rate of 0.001 V/s and the potential range used was from +0.5 V to −0.5 V referred to E_{corr}.

These electrochemical techniques were carried out at several time intervals during 24 h. In order to get a better answer of the corrosion morphology, some samples were exposed during 72 h. After exposure time, selected samples were used to make a superficial analysis through a scanning electron microscope (SEM).

6.4.2 Electrochemical Evaluations

Figure 6.23 shows the results obtained in the corrosion potential (E_{corr}) measurements versus time of the composite and alloy immersed in synthetic seawater at room temperature, atmospheric pressure, and static conditions. E_{corr} values for the alloy and the composite are −0.69 and −0.675 V, respectively.

After 400 s the composite showed a stabilization of corrosion potential. However, the Al-Cu alloy exhibited a heterogeneous behavior even after 1800 s. The difference of the E_{corr} values between composite and alloy is really small; it means that the E_{corr} values are similar in both materials. This behavior should be attributed to the fact that the galvanic current presented in the alloy is similar in the composite, because the electrochemical response of this composite is provided only by the metallic matrix.

Figures 6.24 and 6.25 show the polarization curves of the composite and Al-Cu alloy immersed in seawater, respectively, at room temperature, atmospheric pressure, and static conditions. Figure 6.24 shows that at the beginning of the test (0 h) and in the anodic branch, a transfer charge resistance process is limited to the anodic

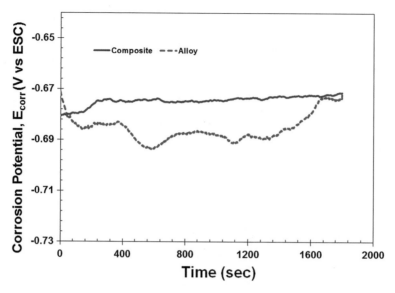

Fig. 6.23 E_{corr} vs time for Al-Cu/TiC composite and Al-Cu alloy immersed in seawater [59]

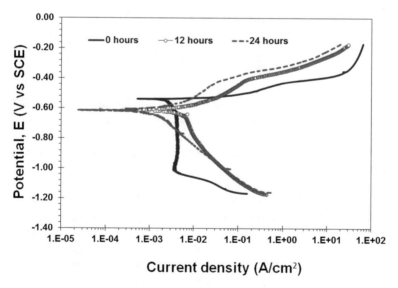

Fig. 6.24 Polarization curves as a function of the time for Al-Cu/TiC composite immersed in seawater [59]

reaction, but at the end of the curve, the anodic Tafel slope increased considerably indicating that a passive corrosion product film is adsorbed on the surface of the composite, whereas in the cathodic branch, a cathodic limiting current density (i_{LIM}) is possible to observe. This i_{LIM} is attributed to the oxygen diffusion process, mainly to the diffusion of the oxygen through the corrosion product film [60].

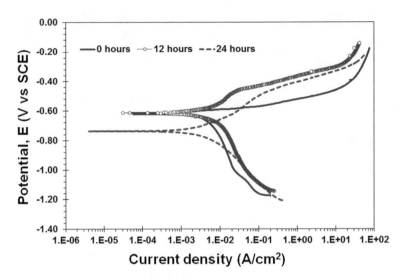

Fig. 6.25 Polarization curves slope as a function of the time for Al-Cu alloy immersed in seawater [59]

Table 6.6 Electrochemical parameter obtained from PCs of composite [59]

Time (h)	ba (V)	bc (V)	I_{corr} (A/cm^2)	CR (mm/year)
0	0.073	1.694	9.31×10^{-6}	0.101
12	0.138	0.353	6.30×10^{-6}	0.068
24	0.168	0.270	2.44×10^{-6}	0.026

At 12 and 24 h of the exposure time, the polarization curves corresponding to anodic branches have slopes that cannot be associated to a pure charge transfer. This feature suggests a contribution of a mass transfer process on the anodic reaction. On the other hand, in the cathodic branches, the reaction remains limited by the oxygen diffusion process.

Table 6.6 presents the electrochemical parameter calculated from these polarization curves for Al-Cu/TiC composite. The parameters are cathodic Tafel slope (bc), anodic Tafel slope (ba), corrosion current density (I_{corr}), and corrosion rate (CR).

It is observe that the highest CR value was obtained at the beginning of the test (0 h). This behavior should be attributed to the fact that at the beginning of the test, the surface of the composite was active. It is important to point out that the CR decreased as the exposure time evolved; it is attributed to the formation of a typical passive corrosion product film on the aluminum.

Figure 6.25 shows the PCs of the polarization curves of the Al-Cu alloy immersed in seawater. Figure 6.25 shows that at the beginning of the test (0 h) and in the anodic branch, a transfer charge resistance process is limited to the anodic reaction, but at the end of the curve, the anodic Tafel slope increased considerably indicating that a passive corrosion product film is adsorbed on the surface of the alloy. On the other hand, in the cathodic branch, a not well-defined cathodic limiting current density (i_{LIM}) can be observed.

When the Al-Cu alloy was exposed to 12 and 24 h, the results of the alloy are similar to the composite (Fig. 6.24); the polarization curves corresponding to anodic branches have slopes that cannot be associated to a pure charge transfer. This feature suggests a contribution of a mass transfer process on the anodic reaction.

Table 6.7 presents the electrochemical parameter calculated from the polarization curves of Fig. 6.25. The values of the Tafel slope presented in this table were used in all calculations of CR.

Table 6.7 shows that the highest CR value was obtained at the middle of the test (12 h). This behavior should be attributed to the fact that the passive corrosion product film formed on the surface of the alloy was broken (highest CR values) by the chloride ion actions, but the CR decreased at 24 h indicating that the passive film was regenerated.

Figure 6.26 shows the corrosion rate in function of the time for Al-Cu alloy and Al-Cu/TiC composite. It is clear the influence of TiC on the Al-Cu alloy on the composite, because the lowest CR values correspond to the alloy sample.

Table 6.7 Electrochemical parameter obtained from PCs of Al-Cu alloy immersed in seawater [59]

Time (h)	ba (V)	bc (V)	I_{corr} (A/cm^2)	CR (mm/year)
0	0.069	0.405	1.46×10^{-6}	0.016
12	0.156	0.448	3.37×10^{-6}	0.037
24	0.131	0.251	2.03×10^{-6}	0.022

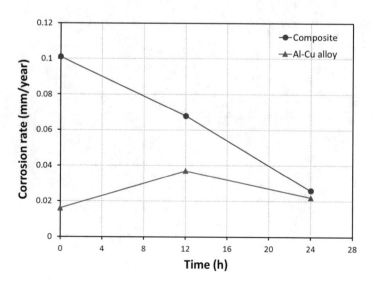

Fig. 6.26 Corrosion rate for Al-Cu/TiC composite and Al-Cu alloy at different exposure times

6.4.3 Surface Analysis by SEM

Figure 6.27 shows the micrographs obtained by SEM from Al-Cu alloy and Al-Cu/TiC composite before and after being exposed to seawater for 72 h.

After being exposed to the Al-Cu alloy (Fig. 6.27b) and the Al-Cu/TiC composite (Fig. 6.27d), it is possible to observe a general corrosion and some zones with localized corrosion, specifically pitting corrosion. It is important to mention that the mechanism that limited this pitting corrosion type is the aeration differential and it was caused by the rupture of corrosion product film (the passive film) on the surface of the metallic alloy.

In summary, according to electrochemical study of the Al-Cu alloy and Al-Cu/TiC composite exposed to seawater (at static conditions, room temperature, and atmospheric pressure), it is possible to conclude that the corrosion rate values of the Al-Cu/TiC composite were higher than the CR values of the Al-Cu alloy; this fact is attributed to the titanium carbide that should induce crevice corrosion. The corrosion of the composite and alloy is a mix process, because a charge transfer resistance

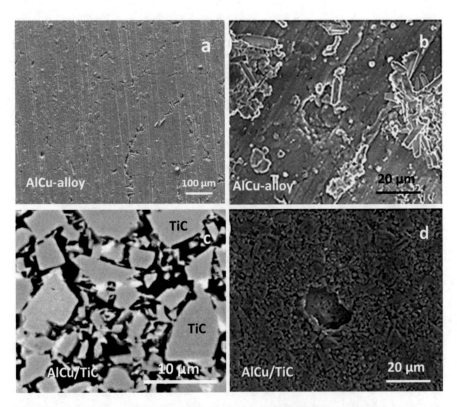

Fig. 6.27 SEM micrographs of (**a–c**) Al-Cu alloy and Al-Cu/TiC composite before exposure to seawater, (**b–d**) Al-Cu alloy and Al-Cu/TiC composite after being exposed to seawater during 72 h [59]

corresponding to a pure corrosion process is limited by a diffusional resistance (diffusion polarization induced by the oxygen diffusion process through of the corrosion product film). The mechanism of the alloy and composite corrosion is a differential aeration cell, where it can produce pitting or crevice corrosion form.

6.5 Corrosion Behavior of Al-Cu-Li/SiC Composites

The corrosion behavior of Al-Cu-Li/SiC$_p$ and Al-Cu/SiC$_p$ composites exposed to NaCl solutions with different pH values was studied [61]. The corrosion behavior of the composites was also compared with the corrosion behavior of pure aluminum. Corrosion potential measurements and potentiodynamic polarization were used to study the corrosion behavior of the composites. The microstructure and morphology of the composites were studied through X-ray diffraction (XRD) and scanning electron microscope (SEM).

The last years Al-Cu-Li/SiC$_p$ and Al-Cu/SiC$_p$ composites have been used for aerospace and aircraft applications. Mechanical properties of Al-Li alloys have been studied extensively, and in general the addition of lithium improves the elastic modulus and the specific stiffness and decreases the density and surface tension of the Al alloys. However little work has been done on the corrosion behavior of metal matrix composites with Al-Cu-Li alloys.

The heterogeneous microstructure in MMC is the cause for its poor corrosion resistance. An unsuitable process of manufacture can be the cause of defects in the interface of particle matrix. The defects found commonly are cracks, lack of adhesion, cavities, and hollows, among others. Some researchers have focused their effort to study and to understand the corrosion behavior [1–3, 13], and others have been dedicated to find strategies to improve the corrosion resistance of the MMC [15, 25–30, 34–36].

Most of the results conclude that addition of reinforcement particles decrease the corrosion resistance of composites in comparison to the corrosion resistance of the matrix [1, 25–30, 62–65]. However, there are some reports where the particles are inert [66–68] or improve the corrosion resistance of the composite material [13, 32].

Hwang et al. [69] found that the metal-ceramic interface is a discontinuity and apparently debilitates the formation of a passive film. Some other problems such as the presence of defects in composites and the different phases involved in the composite could produce galvanic corrosion. This problem is focused in the matrix-reinforcement interface. The additions of alloying elements into the matrix significantly influence the corrosion resistance of the MMC [70–74]. Lithium has been used recently like alloying element in aluminum alloys, with the purpose to increase corrosion resistance. Al-Li alloys are being used more commonly in aeronautical and aerospace applications [75, 76], which in turn has increased the interest in understanding its corrosion behavior. Several studies have demonstrated that additions of lithium enhance the corrosion resistance by the formation of the passive film and its stability [77–79].

Ambat et al. [80] studied the corrosion behavior of two aluminum alloys that contain lithium. The alloys were exposed in solutions with 3.5% NaCl with different pH values. Their results showed that the corrosion current is maximal in extreme values of pH (acid and alkaline) and minimum near to neutral values. Studies made with chloride ions suggest that localized corrosion in Al-Li alloys initiates in the intergranular region [81].

Some researchers found that adding rare earths to Al-Li alloys could increase the corrosion resistance of the Al-Li alloys. It is reported that the rare earths decrease the pit density and stress corrosion cracking (SCC) in the Al-Li alloys [82–84]. Diaz et al. [85] have shown that SiC particles in aluminum MMC can be cathodic sites for the reduction of oxygen. The increase of cathodic sites can be an important factor for the increase of corrosion rate in MMC.

6.5.1   Experimental Conditions

Two different composite materials were used in this study. Al-Cu-Li/SiC$_p$ and Al-Cu/SiC$_p$ composites were obtained from Goodfellow in sheet shape with thickness of 1.6 and 2 mm, respectively. The two composites and pure aluminum were used in the electrochemical studies. The chemical composition of the composites is shown in Table 6.8.

Working electrodes (WE) obtained from the composites with an area of 0.5 cm^2 were used. Potentiodynamic polarization curves were obtained using a glass with standard three-electrode arrangement in NaCl solutions with different pH values. All potentials were measured against an Ag/AgCl reference electrode. Platinum wire as auxiliary electrode was used.

6.5.2   Composite Microstructure

The microstructure of the two composites through scanning electron microscopy (SEM) was observed. Figure 6.28 shows a SEM image of the general surface morphology of the Al-Li-Cu/SiC$_p$ and Al-Cu/SiC$_p$ composites. A more homogeneous distribution of the silicon carbide particles (SiC$_p$) in the metal matrix for the Al-Cu/ESiC$_p$ was observed.

Table 6.8 Chemical composition of material composites studied (wt.%)

Composite	Al	Li	Cu	Mg	Mn	SiC
Al-Cu-Li/SiC$_p$	81	2	1.2	0.8	–	15
Al-Cu/SiC$_p$	78	–	3.3	1.2	0.4	17.8

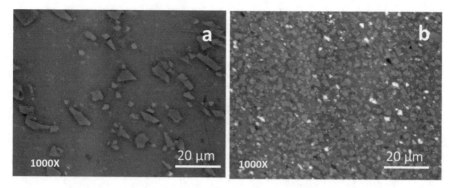

Fig. 6.28 SEM micrograph of composites: (**a**) Al-Li-Cu/SiC$_p$ and (**b**) Al-Cu/SiC$_p$ [61]

6.5.3   Electrochemical Evaluations

Figure 6.29 shows the effect of the pH on the corrosion potential (E_{corr}) at chloride concentration of 0.01, 0.1, and 0.5 M, respectively. At pH 2 and 7, it is observed that the corrosion potential of the composites shift to a noble value with respect to the corrosion potential of aluminum independently of the NaCl concentration. When the Al and composites are exposed to a chloride solution alkaline with pH 12, the corrosion potential changes to more negative values, indicating the high activity of aluminum and the composites. At pH 12 there was no significant difference between the corrosion potential of pure Al and the composites.

Corrosion potential is the characteristic or property of metal and nonmetal surfaces to lose electrons in the presence of an electrolyte. During the process of corrosion, two electrodes are formed spontaneously, a cathode and an anode. The corrosion process involves anodic and cathodic reactions. Both types of reactions are characterized by a Nernst-type equation that describes the relationship between the reversible potential and the ion metal concentration or dissolved species in solution [86]. Therefore, it is expected a strong effect of the alloying element on the corrosion potential. Some studies found that elements like Mg and Zn have a reverse effect, shifting the corrosion potential to active values [87–89]. The corrosion potential can also be modified by defects, such as pinholes, lack of adhesion between matrix and reinforcement, and porosity. The interface between matrix and reinforcement are critical sites for the high corrosion current density in MMC [64, 70–73].

It is observed that the addition of lithium has no significant shift of the corrosion potential. The fluctuations on the corrosion potential of the composites suggest a high activity on the surface that could be associated to localize corrosion.

Figure 6.30 shows the polarization curves for pure Al, Al-Li-Cu/SiC$_p$, and Al-Cu/SiC$_p$ composites as function of the NaCl concentration at pH of 2, 7, and 12, respectively. The corrosion potential of the pure aluminum is more active than the corrosion potential measured for composites at pH 2 and 7. At pH of 12, the behavior is very similar at three NaCl concentration; both aluminum and composites are very active. At pH 2 and 7, the cathodic region of the polarization curves shows a typical cathodic reaction controlled by diffusion.

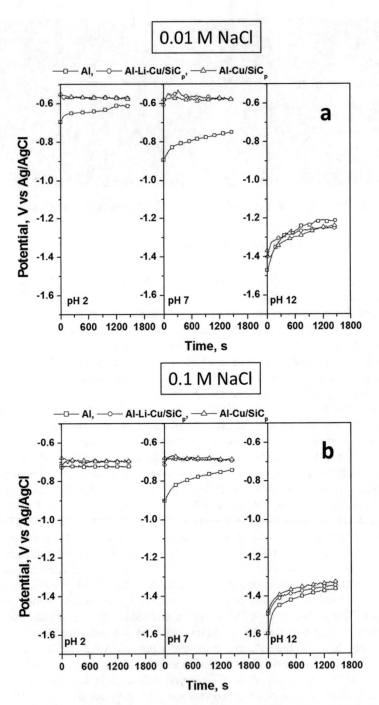

Fig. 6.29 Effect of the pH on the corrosion potential as function of the time for pure Al, Al-Li-Cu/SiC$_p$, and Al-Cu/SiC$_p$ composites exposed to (**a**) 0.01 M NaCl, (**b**) 0.1 M NaCl, (**c**) 0.5 M NaCl solution [61]

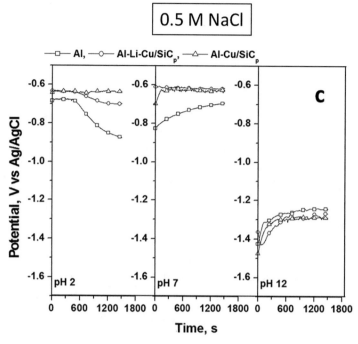

Fig. 6.29 (continued)

The limiting current density on the cathodic site of Fig. 6.30 suggests that the reduction of dissolved oxygen is the predominant reaction. The reason for this is that metal ions cannot be reacted faster than they can arrive at the cathode surface, which is called the limiting current density. It is also observed that cathodic reduction current is higher on the composites compared with cathodic reduction current on pure aluminum. This increase in cathodic sites can be due the presence of silicon carbide particles, which are suitable for the reduction of oxygen [85].

The polarization curve at pH 12 shows that the cathodic Tafel slope change slightly in comparison with the cathodic Tafel slope at pH 2 and 7. The anodic corrosion density is associated with the anodic oxidation reaction of aluminum and the composites. Corrosion normally occurs at a rate determined by an equilibrium between opposing electrochemical reactions. One reaction is the anodic reaction, in which a metal is oxidized, releasing electrons into the metal. The other is the cathodic reaction, in which a solution species (often O_2) is reduced, removing electrons from the metal.

Taking into account the polarization curves obtained in Fig. 6.30, Table 6.9 shows a summary of the corrosion potential and current density in function of the pH and the NaCl concentrations.

Figure 6.31 shows the corrosion rate in function of the pH and NaCl concentration for pure Al, Al-Li-Cu/TiC, and Al-Cu/TiC composites. It is clear the influence

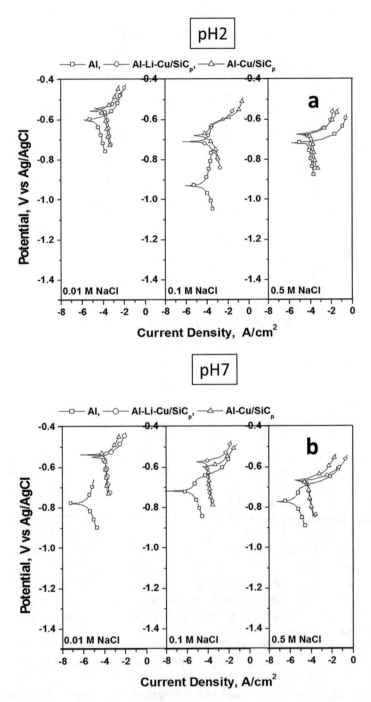

Fig. 6.30 Effect of NaCl concentration (**a–c**) at pH of 2, 7, and 12 on the potentiodynamic polarization curves measured after 1500 s for pure Al, Al-Li-Cu/SiC$_p$, and Al-Cu/SiC$_p$ composites [61]

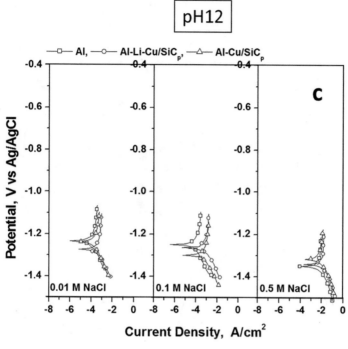

Fig. 6.30 (continued)

Table 6.9 Corrosion potential and current density in function of the pH and the NaCl concentrations

pH	Al		Al-Li-Cu/SiC$_p$		Al-Cu/SiC$_p$	
	E_{corr} (V)	I_{corr} (μA/cm^2)	E_{corr} (V)	I_{corr} (μA/cm^2)	E_{corr} (V)	I_{corr} (μA/cm^2)
0.01 M NaCl						
2	−0.59	3.6	−0.54	14.8	−0.55	27.6
7	−0.75	0.39	−0.54	7.22	−0.54	16.0
12	−1.29	47.0	−1.25	44.5	−1.23	52.9
0.1 M NaCl						
2	−0.91	63.3	−0.69	12.5	−0.69	3.26
7	−0.72	1.4	−0.59	6.6	−0.59	4.2
12	−1.26	18.7	−1.25	198	−1.27	31.0
0.5 M NaCl						
2	−0.77	2.9	−0.67	3.9	−0.67	16.5
7	−0.75	0.6	−0.66	5.9	−0.66	4.2
12	−1.35	864	−1.33	912	−1.31	912

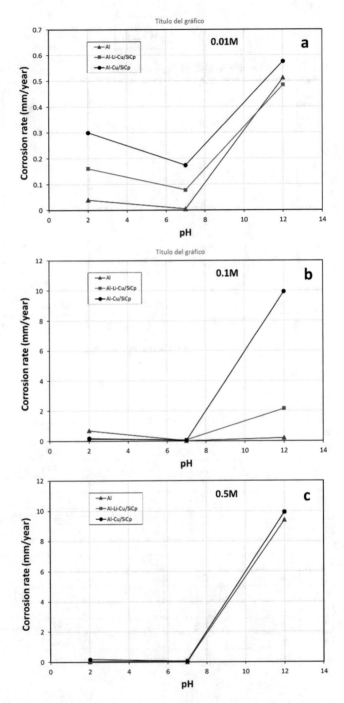

Fig. 6.31 Corrosion rate in function of the pH and the NaCl concentrations: (**a**) 0.01 M, (**b**) 0.1 M, and (**c**) 0.5 M

of TiC particles on the Al and Al-Cu alloy on the composite; in general the higher corrosion rate was exhibited by Al-Cu/SiC composites, and the lowest CR values correspond to pure Al.

At neutral pH it was observed the lowest corrosion rate for all the NaCl concentrations. Meanwhile, at pH 12 was obtained the higher corrosion rate for all the NaCl concentrations.

At moderate NaCl concentration (0.1 M), there is a breakdown potential that can be associated with the pitting potential (E_{pit}). This behavior is more evident at pH 7 and NaCl concentrations of 0.1 and 0.5 M. The electrochemical behavior of the composites was quite similar between them. There was no observed significant effect of Li on the composite corrosion. However the addition of lithium improves the mechanical properties of the Al-Li alloys and reduces significantly (more than Mg) the density and surface tension of the Al [90–92].

In summary, with pH 2 and 7, the corrosion potential of aluminum was more active in comparison with the corrosion potential of composites under the same conditions. Composites exhibit higher corrosion current density than pure aluminum, which could be attributed to the presence of SiC particles that act as cathodic sites. At 0.5 M NaCl solution with pH 12 was the more aggressive for the composites and pure aluminum. The addition of Li into the Al-Cu/SiC$_p$ composite did not affect considerably the corrosion potential neither current density of the composites.

References

1. Albiter A, Contreras A, Salazar M, Gonzalez JG (2006) Corrosion behaviour of aluminium metal matrix composites reinforced with TiC processed by pressureless melt infiltration. J Appl Electrochem 36:303–308
2. Hihara LH, Latanision RM (1994) Corrosion of metal matrix composites. J Int Mater Rev 39 (6):245–264
3. Turnbull A (1992) Review of corrosion studies on aluminium metal matrix composites. Br Corros J 27(1):27–35
4. Makar GL, Kruger J (1993) Corrosion of magnesium. Int Mater Rev 38:138–153
5. Gusieva K, Davies CHJ, Scully JR, Birbilis N (2015) Corrosion of magnesium alloys: the role of alloying. Int Mater Rev 38:138–153
6. Melchers RE (2015) Bi-modal trends in the long-term corrosion of copper and high copper alloys. Corros Sci 95:51–61
7. Leon CA, Lopez VH, Bedolla E, Drew RAL (2002) Wettability of TiC by commercial aluminum alloys. J Mater Sci 37:3509–3514
8. Contreras A, Leon CA, Drew RAL, Bedolla E (2003) Wettability and spreading kinetics of Al and Mg on TiC. Scr Mater 48:1625–1630
9. Contreras A, Albiter A, Perez R (2004) Microstructural properties of the Al-Mgx/TiC composites obtained by infiltration techniques. J Phys 16:S2241–S2249
10. Contreras A, Angeles-Chávez C, Flores O, Perez R (2007) Structural, morphological and interfacial characterization of Al–Mg/TiC composites. Mater Charact 58:685–693
11. Contreras A, Bedolla E, Perez R (2004) Interfacial phenomena in wettability of TiC by Al–Mg alloys. Acta Mater 52:985–994
12. Deuis RL, Green L, Subramanian C, Yellup JM (1997) Influence of the reinforcement phase on the corrosion of aluminium composite coatings. Corrosion 16:440–444

13. Deuis RL, Green L, Subramanian C, Yellup JM (1997) Corrosion behavior of aluminum composite coatings. Corrosion 53(11):880–890
14. Nunes PCR, Ramanathan LV (1995) Corrosion behavior of alumina-aluminum and silicon carbide-aluminum metal-matrix composites. Corrosion 51(8):610–617
15. Shimizu Y, Nishimura T, Matsushima I (1995) Corrosion resistance of Al-based metal matrix composites. Mater Sci Eng A 198:113–118
16. Yao HY, Zhu RZ (1998) Interfacial preferential dissolution on silicon carbide particulate/aluminum composites. Corrosion 54(7):499–507
17. Paciej RC, Agarwala VS (1988) Influence of processing variables on the corrosion susceptibility of metal-matrix composites. Corrosion 44(10):680–684
18. Sun H, Koo EY, Wheat HG (1991) Corrosion behavior of SiC$_p$/6061 Al metal matrix composites. Corrosion 47(10):741–753
19. Trzaskoma P (1990) Pit morphology of aluminum alloy and silicon carbide/aluminum alloy metal matrix composites. Corrosion 46(5):402–409
20. Hihara LH, Latanision RM (1992) Galvanic corrosion of aluminum-matrix composites. Corrosion 48:546–552
21. Modi OP, Saxena M, Prasad BK, Jha AK, Das S, Yegneswaran AH (1998) Role of alloy matrix and dispersoid on corrosion behavior of cast aluminum alloy composites. Corrosion 54 (2):129–134
22. Contreras A, Salazar M, León CA, Drew RAL, Bedolla E (2000) The kinetic study of the infiltration of aluminum alloys into TiC. Mater Manuf Process 15(2):163–182
23. Stearn M, Geary AL (1958) The mechanism of passivating type inhibitors. J Electrochem Soc 105:638–647
24. Albiter A, Contreras A, Bedolla E, Perez R (2003) Structural and chemical characterization of precipitates in Al2024/TiC composites. Compos Part A 34:17–24
25. Candan S (2009) An investigation on corrosion behaviour of pressure infiltrated Al-Mg alloy/SiC composites. Corros Sci 51(6):1392–1398
26. Candan S (2004) Effect of SiC particle size on corrosion behavior of pressure infiltrated Al matrix composites in a NaCl solution. Mater Lett 58:3601–3605
27. Ahmad Z, Abdul Aleem BJ (2002) Degradation of aluminum metal matrix composites in salt water and its control. Mater Des 23(2):173–180
28. Chen C, Mansfeld F (1997) Corrosion protection of an Al 6092/SiC metal matrix composite. Corros Sci 39(6):1075–1082
29. Kiourtsidis GE, Skolianos SM, Pavlidou EG (1999) A study on pitting behaviour of AA2024/SiC$_{(p)}$ composites using the double cycle polarization technique. Corros Sci 41(6):1185–1203
30. Bedolla E, Lemus-Ruiz J, Contreras A (2012) Synthesis and characterization of Mg-AZ91/AlN composites. Mater Des 38:91–98
31. Reyes A, Bedolla E, Perez R, Contreras A (2012) Effect of heat treatment on the mechanical and microstructural characterization of Mg-AZ91E/TiC composites. Compos Interfaces 24:1–17
32. Falcon LA, Bedolla E, Lemus J, Leon CA, Rosales I, Gonzalez-Rodriguez JG (2011) Corrosion behavior of Mg-Al/TiC composites in NaCl solution. Int J Corros 2011:1–7
33. Pardo A, Merino MC, Coy AE, Arrabal R, Viejo F, Matykina E (2008) Corrosion behaviour of magnesium/aluminium alloys in 3.5 wt.% NaCl. Corros Sci 50(3):823–834
34. Nunez-Lopez CA, Skeldon P, Thompson GE, Lyon P, Karimzadeh H, Wilks TE (1995) The corrosion behaviour of Mg alloy ZC71/SiC$_p$ metal matrix composite. Corros Sci 37(5):689–708
35. Suqiu J, Shusheng J, Guangping S, Jun Y (2005) The corrosion behaviour of Mg alloy AZ91D/TiC$_p$ metal matrix composite. Mater Sci Forum 488–489:705–708
36. Tiwari S, Balasubramaniam R, Gupta M (2007) Corrosion behavior of SiC reinforced magnesium composites. Corros Sci 49(2):711–725
37. Salman SA, Ichino R, Okido M (2010) A comparative electrochemical study of AZ31 and AZ91 magnesium alloys. Int J Corros 2010:1–7
38. Singh IB, Singh M, Das S (2015) A comparative corrosion behavior of Mg, AZ31 and AZ91 alloys in 3.5% NaCl solution. J Magnes Alloys 3:142–148

39. Budruk AS, Balasubramaniam R, Gupta M (2008) Corrosion behaviour of Mg-Cu and Mg-Mo composites in 3.5% NaCl. Corros Sci 50(9):2423–2428
40. Huang HH, Tsai WT, Lee JT (1996) Electrochemical behavior of A516 carbon steel in solutions containing hydrogen sulfide. Corrosion 52(9):708–716
41. Ungaro ML, Carranza RM, Rodriguez MA (2012) Crevice corrosion study on alloy 22 by electrochemical noise technique. Proc Mater Sci 1:222–229
42. Cottis RA (2001) Interpretation of electrochemical noise data. Corrosion 57:265–285
43. Cowan KG, Harrison JA (1980) The automation of electrode kinetics—III. The dissolution of Mg in Cl^-, F^- and OH^- containing aqueous solutions. Electrochim Acta 25(7):899–912
44. Harris SJ, Noble B, Trowsdale AJ (1996) Corrosion behaviour of aluminium matrix composites containing silicon carbide particles. Mater Sci Forum 217–222:1571–1579
45. Duran-Olvera JM (2017) Análisis electroquímico del proceso de corrosión del composito TiC-Ni en agua de mar sintética. Thesis, Universidad Veracruzana, México
46. Duran-Olvera JM, Orozco-Cruz R, Galván-Martínez R, León CA, Contreras A (2017) Characterization of TiC/Ni composite immersed in synthetic seawater. MRS Adv 2(50):2865–2873
47. Bhattacharyya JJ, Mitra R (2012) Effect of hot rolling temperature and thermal cycling on creep and damage behavior of powder metallurgy processed Al–SiC particulate composite. Mater Sci Eng 557:92–105
48. Kala H, Mer KKS, Kumar S (2014) A review on mechanical and tribological behaviors of stir cast aluminum matrix composites. Proc Mater Sci 6:1951–1960
49. Karbalaei-Akbari M, Rajabi S, Shirvanimoghaddam K, Baharvandi HR (2015) Wear and friction behavior of nanosized TiB_2 and TiO_2 particle-reinforced casting A356 aluminum nanocomposites: a comparative study focusing on particle capture in matrix. J Compos Mater 49(29):3665–3681
50. Leon CA, Arroyo Y, Bedolla E (2006) Properties of AlN-based magnesium-matrix composites produced by pressureless infiltration. Mater Sci Forum 502:105–110
51. Silverman DC (2003) Aqueous corrosion, corrosion: fundamentals, testing and protection. In: ASM handbook, vol 13A. ASM International, Materials Park, Ohio
52. ASTM D1141 Standard practice for the preparation of substitute ocean water (2013)
53. ASTM G5 Standard reference test method for making potentiostatic and potentiodynamic anodic polarization measurements (2014)
54. Bastos Segura JA (2000) Comportamiento electroquímico del níquel en una matriz de resina epoxidica. Doctoral dissertation, Universitat de Valencia
55. Zamin M, Ivés MB (1973) Effect of chloride ion concentration on the anodic dissolution behavior of nickel. Corrosion 29:319–324
56. Real SG, Barbosa MR, Vilche JR, Arvía AJ (1990) Influence of chloride concentration on the active dissolution and passivation of nickel electrodes in acid sulfate solutions. J Electrochem Soc 137:1696–1702
57. Jones DA (1996) Principles and prevention of corrosion, 2nd edn. Prentice-Hall, Upper Saddle River, pp 1–108, 146–150, 368–370
58. ASTM G1 standard practice for preparing, cleaning, and evaluation corrosion test specimens (2011)
59. Alvarez-Lemus N, Leon CA, Contreras A, Orozco-Cruz R, Galvan-Martinez R (2015) Chapter 15: electrochemical characterization of the aluminum–copper composite material reinforced with titanium carbide immersed in seawater. In: Perez R, Contreras A, Esparza R (eds) Materials characterization. Springer, Cham, pp 147–156
60. Galvan-Martinez R, Cabrera D, Galicia G, Orozco R, Contreras A (2013) Electrochemical characterization of the structural metals immersed in natural seawater: "in situ" measures. Mater Sci Forum 755:119–124
61. Lugo-Quintal J, Díaz-Ballote L, Veleva L, Contreras A (2009) Effect of Li on the corrosion behavior of Al-Cu/SiCp composites. Adv Mater Res 68:133–144
62. Abdallah M, Omar AA, Kandil A (2003) Production and corrosion behaviour of A7475 and Sicp. Bull Electrochem 19:405–412

63. Singh N, Vadera KK, Kumar AVR, Singh RS, Monga SS, Mathur GN (1999) Corrosion behaviour of 2124 aluminium alloy-silicon carbide metal matrix composites in sodium chloride environment. Bull Electrochem 15:120–123

64. Bhat MSN, Surappa MK, Nayak HVS (1991) Corrosion behaviour of silicon carbide particle reinforced 6061/Al alloy composites. J Mater Sci 26(18):4991–4996

65. Sun H, Koo EY, Wheat HG (1991) Interfacial preferential dissolution on silicon carbide particulate/aluminum composites. Corrosion 47(9):741–749

66. Rohatgi PK, Xiang CH, Gupta N (2018) Aqueous corrosion of metal matrix composites. Mater Sci Eng 4:287–312

67. Contreras A, Lopez VH, Bedolla E (2004) Mg/TiC composites manufactured by pressureless melt infiltration. Scr Mater 51:249–253

68. Kolman DG, Butt DP (1997) Corrosion behavior of a novel SiC/Al₂O₃/Al composite exposed to chloride environments. J Electrochem Soc 144:3785–3791

69. Hwang WS, Kim HW (2002) Galvanic coupling effect on corrosion behavior of Al alloy-matrix composites. Met Mater Int 8:571–575

70. Pardo A, Merino MC, Arrabal R, Feliu S, Viejo F, Carboneras M (2005) Enhanced corrosion resistance of A3xx.x/SiCₚ composites in chloride media by La surface treatments. Electrochim Acta 51:4367–4378

71. Pardo A, Merino MC, Arrabal R, Merino S, Viejo F, Carboneras M (2006) Effect of Ce surface treatments on corrosion resistance of A3xx.x/SiCₚ composites in salt fog. Surf Coat Technol 200:2938–2947

72. Pardo A, Merino S, Merino MC, Barroso I, Mohedano M, Arrabal R, Viejo F (2009) Corrosion behaviour of silicon carbide particle reinforced AZ92 magnesium alloy. Corros Sci 51:841–849

73. Pardo A, Merino MC, Arrabal R, Feliu S (2007) Effect of La surface coatings on oxidation behavior of aluminum alloy/SiCₚ composites. Oxid Met 67:6786

74. Datta J, Datta S, Banerjee MK, Bandyopadhyay S (2004) Beneficial effect of scandium addition on the corrosion behavior of Al–Si–Mg–SiCₚ metal matrix composites. Compos Part A 35:1003–1008

75. Staley JT, Lege DJ (1993) Advances in aluminium alloy products for structural applications in transportation. J Phys Colloq 3:C7-179–C7-190

76. Rao KTV, Ritchie RO (1998) High-temperature fracture and fatigue resistance of a ductile β-TiNb reinforced γ-TiAl intermetallic composite. Acta Mater 46(12):4167–4180

77. Roper GW, Attwood PA (1995) Corrosion behaviour of aluminium matrix composites. J Mater Sci 30:898–903

78. Murthy KSN, Dwarakadasa ES (1995) Role of Li⁺ ions in corrosion behaviour of 8090 Al–Li alloy and aluminium in pH 12 aqueous solutions. Br Corros J 30:111–115

79. Salghi R, Bazzi L, Zaafrani M (2003) Effet d'ínhibition de la corrosión de deux alliages d'aluminium 6063 et 3003 par quelques cations metallique en milieu chlorure. Acta Chim Slov 50:491–495

80. Ambat R, Dwarakadasa ED (1992) The influence of pH on the corrosion of medium strength aerospace alloys 8090, 2091 and 2014. Corros Sci 33:681–690

81. Damborenea JJ, Conde A (2000) Intergranular corrosion of 8090 Al–Li: interpretation by electrochemical impedance spectroscopy. Br Corros J 35:48–53

82. Davo B, Damborenea JJ (2004) Corrosión e inhibición en aleaciones de aluminio de media resistencia. Rev Metal 40:442–446

83. Davo B, Damborenea JJ (2004) Use of rare earth salts as electrochemical corrosion inhibitors for an Al–Li–Cu (8090) alloy in 3.56% NaCl. Electrochim Acta 49:4957–4965

84. Davo B, Conde A, Damborenea JJ (2005) Inhibition of stress corrosion cracking of alloy AA8090 T-8171 by addition of rare earth salts. Corros Sci 47:1227–1237

85. Diaz-Ballote L, Veleva L, Pech-Canul MA, Pech-Canul MI, Wipf DO (2004) Activity of SiC particles in Al-based metal matrix composites revealed by SECM. J Electrochem Soc 151: B299–B303

86. Bard AJ, Faulkner LR (2001) Electrochemical methods: fundamentals and applications, 2nd edn. Wiley, New York
87. Baldwin KR, Bates RI, Arnell RD, Smith CJE (1996) Aluminium-magnesium alloys as corrosion resistant coatings for steel. Corros Sci 38:155–170
88. Kim Y, Buchheit RG (2007) A characterization of the inhibiting effect of Cu on metastable pitting in dilute Al–Cu solid solution alloys. Electrochim Acta 52:2437–2446
89. Ralston KD, Birbilis N, Cavanaugh MK, Weyland M, Muddle BC, Marceau RKW (2010) Role of nanostructure in pitting of Al–Cu–Mg alloys. Electrochim Acta 55:7834–7842
90. Sankaran KK, Grant NJ (1980) The structure and properties of splat-quenched aluminum alloy 2024 containing lithium additions. Mater Sci Eng 44:213–227
91. Hatch JE (1984) Aluminum properties and physical metallurgy. American Society for Metals, Materials Park, Ohio
92. Garrard WN (1994) Corrosion behavior of aluminum-lithium alloys. Corrosion 50(3):215–225

Chapter 7
Wear of Composites

7.1 Introduction

The word tribology is derived from the Greek word "tribos" which means rubbing or friction and is known as the science and technology that studies the movements of surfaces in contact when they are in relative motion and involves the friction caused between these surfaces, as well as wear and lubrication. The renaissance artist Leonardo da Vinci was the first to propose basic friction laws as the proportionality between the normal force and limiting friction force, from which the concept of coefficient of friction comes off. Da Vinci deduced the laws governing the motion of a rectangular block sliding on a flat surface, although his notes were never published. Even with the importance of the wear subject, little was done for many years. Of particular note are the proposals of the physicist Guillaume Amontons in 1699 who publishes the first rules of friction derived from the study of the sliding contact of two flat bodies, as well as the publication of Osborne Reynolds in 1866 about hydrodynamic lubrication, from which the mathematical theory of lubrication arises and is currently used.

In recent decades, governments and companies in industrialized countries have recognized the importance of reducing energy consumption that is used to overcome friction and wear in industrial processes and transportation. As a result, programs aimed to reduce energy consumption and the costs of friction and wear in industrial processes have been stablished. One of earliest studies was conducted in 1966 by Jost [1] commissioned by the British government. The study concludes that by applying tribological principles great savings could be achieved. According to the report, in 1974 the British industry could have conservatively saved at least £ 100 million p.a. (USD$ 230 million). On the other hand, Pinkus and Wilcock [2] carried out in 1977 a study to determine the potential of energy conservation, applying research and development programs in tribology, oriented to equipment used in ground transportation, turbomachinery, and industrial machinery. From the study carried out by these researchers, it was concluded that savings of up to

© Springer Nature Switzerland AG 2018
A. Contreras Cuevas et al., *Metal Matrix Composites*,
https://doi.org/10.1007/978-3-319-91854-9_7

one-tenth of the total energy consumption in the United States could have been achieved if the recommended R&D plan and its findings were implemented.

There is great concern about pollution caused by industry and means of transportation, in particular by the increase of carbon dioxide emissions, the reason why it is very important to reduce CO_2 emissions and reduce consumption of fuel from fossil sources, which can be achieved by developing lighter transport vehicles and reducing expended energy in overcoming friction in industrial machines and means of transport.

Holmberg et al. [3] performed calculations about global fuel consumption used to overcome friction in passenger cars, taking into account friction in machine, transmission, tires, and brakes. Concluding that direct friction losses of 28% were obtained if the braking system is not taken into account. The world used 208 billion liters of fuel to overcome friction in passenger cars. Another conclusion is that, if new technologies were taken into account to reduce friction in passenger cars, the liters of fuel could be reduced by 18% in a short time (5–10 years) and by 61% in a term between 15 and 20 years, which would be equivalent to an energy saving of 117 billion and 385 billion liters, respectively, and a very good contribution in pollution reduction, since it would reduce emission of CO_2 in 290 and 960 million tons, respectively.

7.2   Friction and Coefficient of Friction

The concept of friction was first mentioned in the late fifteenth century by Leonardo da Vinci and is defined as the force that opposes movement of one body over another. There are two types of relative movements: sliding and rolling; in both cases a tangential force (F) is required to move one body over another from the rest; the relation between friction force and normal applied load (W) is known as coefficient of friction (μ), whose mathematical definition is:

$$\mu = F/W \tag{7.1}$$

where F is the frictional force, W is the normal load on the contact area, and μ is the coefficient of friction.

The coefficient of friction (COF) can be a good reference to know wear in tribological systems because generally the lower coefficient of friction the less wear, although in some cases this is not real. The coefficient of friction in metals and alloys depends largely on normal load, speed, composition of materials in contact, microstructure, and conditions under which tests are performed. Environment influences determination of μ; meanwhile, when tests are performed in air or vacuum, there are marked differences; since in vacuum there is low oxygen potential (Po_2), which hinders the formation of oxides that could be formed between the bodies in contact, acting as lubricant and forming a barrier to avoid direct contact between bodies in motion, thus reducing the coefficient of friction. Temperature is also an important factor because at certain temperatures phase changes in solid state could occur or induce formation of oxides; environment also influences the COF because of moisture, and water or gases present in the tribological pair can modify surfaces in contact and therefore coefficient of friction.

7.3 Wear of Materials

Wear is the erosion suffered by one solid surface by the action of another. It is related to interactions between surfaces and more specifically to the removal of material from a surface as a result of mechanical action [4]. Although it is very difficult to measure wear in laboratory and project it at industrial level, there are different methods that can be used, since in reality there are diverse types of wear (rolling, oscillation, impact, solid particle erosion, etc.) and circumstances in which wear process becomes a problem. A number of standardized methods have been proposed for each type of wear, although each has its limitations; some techniques have received a good acceptance at laboratory level worldwide. Even if similar techniques are used, the obtained results by different researchers show divergences, since both conditions and procedures for carrying out tests are generally different. Therefore, it is advisable to know precisely the conditions used by other researchers in order to make appropriate comparisons. In case of industrial scaling, it is necessary to select both, method and conditions of the tribosystem at laboratory level and the closest to the practice. Although the best way to test wear resistance of any material is under real processes, this is slow and costly; therefore different test devices currently accepted are used. Laboratory tests used to measure sliding wear are pin-on-disc, block-on-disc, pin-on-ring, and reciprocating or pin-on-flat.

At laboratory level, different wear testing apparatus are used to study sliding wear. Some of the most common are presented in Fig. 7.1. Figure 7.1a shows the pin-on-disc method, where the pin usually presents various geometries and remains static, while the disc (counterpart) rotates at selected speed; on the other hand, in the pin-on-ring, the pin is placed by exerting pressure perpendicularly to the ring, as is shown in Fig. 7.2b; in the case of block-on-disc showed in Fig. 7.1c, the block-shaped specimen is placed perpendicular to one side of the rotating disc; meanwhile in the reciprocating or pin-on-flat (Fig. 7.1d), the specimen moves back and forth at high frequencies, while the pin remains static by exerting the predetermined load, or the pin is moving while the specimen remains static; the pin may have a flat or rounded surface.

The wear of the materials is usually calculated using the Archard equation [5]:

$$Q = KW/H \tag{7.2}$$

where Q is the worn volume per distance, W is the applied standard load, H is the hardness of the softest material, and K is a constant.

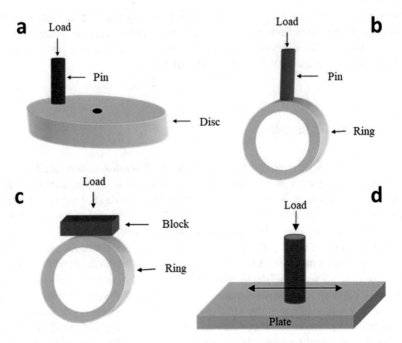

Fig. 7.1 Schematic illustration of geometries used in sliding wear tests: (**a**) pin-on-disc, (**b**) pin-on ring, (**c**) block-on-ring, and (**d**) reciprocating or pin-on-flat

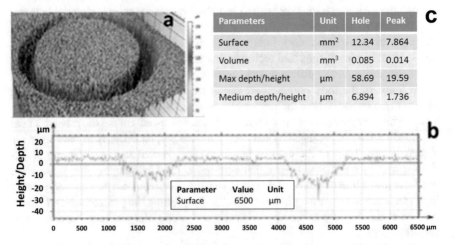

Fig. 7.2 Wear footprint in sliding wear: (**a**) 3D schema, (**b**) typical profile, and (**c**) data table

For engineering applications, the ratio $K/H = k$ is more useful, since it can be applied in different types of materials, provides a measure of deterioration severity, and allows comparing results with other researchers when using different loads and distances or times in the test [6]:

$$k = V/WL \qquad (7.3)$$

where k is the wear rate or specific wear rate in mm^3/Nm, W the normal applied load, L the distance traveled in meters, and V the worn volume in mm^3 [7].

Worn volume can be determined using different methods; weighting loss material is perhaps the most used, recording the mass of test pieces before and after the test; however this is accurate only when high amounts of lost material are present; for a long time, the most used method to calculate wear material due to its greater precision has been measuring wear footprint profiles. In this method longitudinal and transverse profiles of worn footprint are obtained, and using different calculation methods, the volume of track wear is determined. Recently, new equipment capable of scanning the whole worn track with optical devices (together with adequate software) that determine the total volume in a highly precise technique has been developed; in addition, it is possible to generate images of wear footprints, as is shown in Fig. 7.2 the typical wear profile of a wear footprint.

7.4 Tribosystems

A tribosystem is identified as a tribological pair of great utility to describe attrition processes with the purpose of transforming input and output data in technologically useful information. Input and output data can be movement, work, material, and information; on the other hand, unwanted inputs are vibration, heat, dirt, and reactive environment. Friction and wear induce undesirable outlets, such as wear, heat, vibration, and noise fragments. A tribosystem is generally formed by four components: solid body, counterpart, interface, and environment. A scheme of a tribosystem is presented in a simplified way in Fig. 7.3.

7.5 Wear Mechanisms

Although some researchers suggest different classifications of wear mechanisms, the common feature is the use of these mechanisms to distinguish wear processes; DIN 50320 proposes the following mechanisms—adhesion, abrasion, surface fatigue, and tribochemical reaction—being these the most used nowadays; however, researchers propose variants such as corrosive wear, erosion, impact, cavitation wear, softening, and oscillation wear, among others. Regardless of the tests in which material is prepared, it will present at least one wear mechanism or a

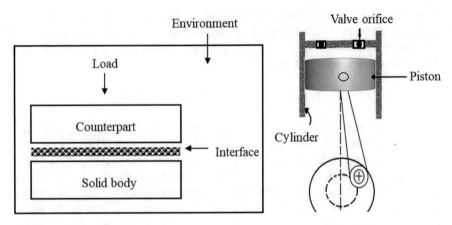

Fig. 7.3 Scheme of a tribosystem

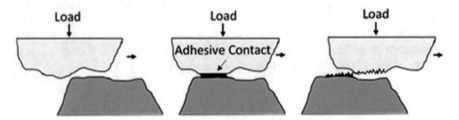

Fig. 7.4 Scheme of adhesion

combination of more than one. The knowledge of wear mechanisms is very important, since they help to predict the variables that must be used to control severe wear,
this by changing parameters, such as normal load, sliding speed, amount, and shape
of reinforcement among other.

Adhesion is the formation and detachment of adhesive bonds at interface; it
occurs when surfaces in contact adhere between each other for an instant and the
driving force separates them violently but in a different area from that of the joint
generating a loss of material in one of contacting parts. Wear by this mechanism
grows with subsequent adhesions and peeling in the remaining area of surfaces in
contact. Figure 7.4 presents schematically an adhesive mechanism. Adhesion in
tribological couples is more susceptible if surfaces in contact are highly clean—
there is more contact area between them—and with high solubility between bodies in
contact. In addition, temperature influences this mechanism, because smooth surfaces and increasing contact area can induce diffusion at interface or generate oxides
that would act as a barrier inhibiting diffusion. The most accepted theories about
adhesion are mechanical coupling, diffusion, electronic, and adsorption theories.

Abrasion is the removal of material due to scratches because of the presence of
hard particles in one of two moving surfaces that scratch the other surface; this is
cyclically repeated, deforming surfaces in contact until material is detached by

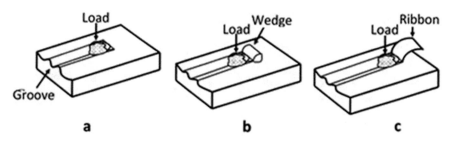

Fig. 7.5 Scheme of abrasion: (**a**) plowing, (**b**) wedge formation, and (**c**) cutting

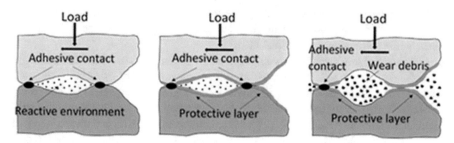

Fig. 7.6 Scheme of tribochemical mechanism

fatigue or pure cut. Abrasive mechanism can be classified in two or three bodies of abrasion; in this last, the abrasive particles are loose between bodies in contact, whereas in the two bodies they are fixed; for that reason abrasion of three bodies is smaller by one or two orders of magnitude with respect to that of two bodies. In this wear mechanism at least three forms of deformation are recognized: plowing, cutting, and wedge formation [8, 9]. Figure 7.5 shows schemes of abrasion forms. In case of plowing, material displaced by the abrasive particles forms crests on groove side which after several cycles are displaced by fatigue; in case of wedge formation, abrasive particle forms grooves that push material toward the front in a crest and others toward the sides; the form of cutting occurs when the attack angle in abrasive particles reaches a critical value, and the wear occurs by plowing a groove and pushing material out in the form of ribbons.

 Tribochemical reaction is the formation of chemical products as a result of chemical interactions caused by environment and temperature, generated by rubbing of bodies in contact in a tribosystem. Figure 7.6 present schematically a tribochemical mechanism. The wear process occurs by a continuous removal of reaction layer and the formation of a new one on surfaces in contact. Load and sliding speed influence tribochemical wear, as high loads and speed increase the temperature of tribosystems, which favors chemical reactions and products such as oxides of metals in contact. Mechanical properties of layer formed at interface determine the resistance to be removed; this layer can be thick but generally is fragile and at the moment of cracking produces abrasive fragments. Nevertheless thin and hard layers can also generate hard particles, with the increase in the load that

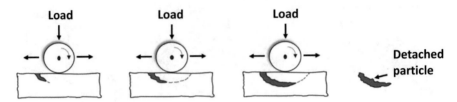

Fig. 7.7 Scheme of surface fatigue

could contribute to abrasive wear. A layer with good mechanical properties inhibits adhesive contacts that occur initially between the rough asperities. In some cases, the formation of some tribochemical products at interface promotes the reduction of wear while increasing normal load [10].

Surface fatigue, wear by fatigue, is caused by cycles of loads on the surface of the material, causing cracks that appear on both surface and subsurface; this anomaly occurs when roughness of both materials are in contact accompanied by high local stresses. Materials containing hard inclusions or other imperfections that act as void generation nucleus under plastic deformation, where cracks start and may be parallel to surface but below it, after many cycles, cracks reach the surface generating thin and elongated sheets, leaving small holes aligned on the surface, this theory known as delamination was introduced by Suh [11]. This wear mechanism occurred in MMCs especially when high normal loads are applied [12], and these are manufactured with matrices softer than hard reinforcements such as ceramic and intermetallic. Figure 7.7 presents a scheme of surface fatigue mechanism.

In most wear processes, more than one wear mechanism is present. In first stages, only one of them is present, and as the contact surfaces change, other wear mechanisms appear that can act at the same time with others, but at the end of the test, the dominant wear mechanism can be identified, which determines the volume of worn material [13].

Generally wear resistance is greater when the coefficient of friction between moving bodies is low, so it is important to reduce friction between tribological pair, although in some cases, under certain normal load ranges, this does not happen [14]; that is, some materials with higher coefficient of friction present higher wear resistance than those with lower coefficient of friction.

The Archard equation (7.2) states that wear is proportional to normal load and inversely proportional to material hardness and it is used in most material wear process; however, some materials do not behave in accordance to this law, at least under certain normal load ranges [10]. Likewise and in agreement with Archard equation, most materials register greater resistance to wear with greater hardness, mainly in monolithic materials like steel and a variety of alloys of great interest in industry. However, composite materials and some iron foundries do not behave in the same way, as there are various factors that influence this behavior. Distance and sliding speed also play a very important role in wear tests, since depending on speed and sliding distances, phenomena such as temperature increase and formation of tribochemical products on interfaces or phase transformations in tribosystem change initial conditions.

7.6 Wear on Metal Matrix Composites

As in metals and their alloys, friction and wear in metal matrix composites depend on many factors. This section will be dedicated to review different works about wear of MMC, especially composites with light alloys reinforced with particles, fibers, and whiskers. As mentioned previously, there are a great variety of factors that influence wear processes in MMC, such as normal applied load, sliding speed, time or distance, quantity and shape of reinforcements, and MMC processing, among others.

7.6.1 Influence of Normal Applied Load

Normal load is one parameter that most influences have in common; friction and wear processes of MMC and most researchers agree that the higher load, the greater wear. Nevertheless, normal load in certain range and conditions reduces wear in MMC [10, 15]. Table 7.1 shows manufacturing processes, matrices, and reinforcements, as well as conditions under which some researchers performed wear tests. Ramakoteswara et al. [12] investigated wear behavior in composites made of aluminum alloy reinforced with TiC particles; they found that reinforcement reduces wear compared to matrix but up to 8 vol.% TiC; nevertheless, lower values of coefficient of friction in the MMC than the alloy were reported, attributing this effect to the formation of iron-rich phase layers formed by iron detached from counterpart. On the other hand, wear rate increases in both the alloy and MMC by increasing normal load. Lim et al. [13] investigated an MMC made with a magnesium and aluminum alloy reinforced by SiC particles. They report that under 10N load, wear resistance of composites was higher at all tested speeds, except at 5 m/s. However, at 30N load, the matrix exhibits higher wear resistance than composite, except at speeds of 1 and 2 m/s, and concludes that even when several wear mechanisms occurred, the dominant was the oxidative mechanism at lower loads. Selvam et al. [16] investigated the behavior of a composite made of magnesium matrix reinforced with zinc oxide nanoparticles (ZnO). They conclude that the addition of ZnO nanoparticles into the matrix improves wear resistance, and wear rate increases with the increase of applied normal load using the three sliding speeds studied, attributing this to the abrasion mechanism, particularly because of the plowing and large debris trapped between pin and disc. Additionally the coefficient of friction decreases with applied load and sliding speed.

Falcón et al. [17] studied the wear behavior of composites made with AZ91E magnesium alloy reinforced with 50% vol TiC_p. The composites were worn against

Table 7.1 Research about load influence in wear

Ref.	Sliding speed (ms⁻¹)	Normal load (N)	Apparatus	Reinf. type	Reinf. size (µm)	Matrix type	Fabrication	Counterpart type
[10]	8 Hz*	2, 4, 6, 8, 10	Ball-on-flat	SiC, 9.8, 26.3 vol.%	25	Mg alloy	DMD	Steel balls SAE 52100
[12]	2	10, 20, 30	Pin-on-disc	TiC$_p$ 2, 4, 6, 8, 10 wt.%	2	AA7075	SC	EN 32 steel
[13]	0.2, 0.5, 1.2, 5	10, 30	Pin-on-disc	SiC$_p$ 8 vol.%	14	Mg-Al alloy	PM	AISI-O1 tool steel
[15]	5, 7, 9	50, 80	Pin-on-disc	SiC$_p$ 20 wt.%; SiC 10, Gr 6, Sb$_2$S$_3$ 3 wt.%	SiC 34; Gr 45	Al alloy	PM	EN24 steel
[17]	200 rpm*	0.5, 1.0 MPa*	Pin-on-disc	TiC$_p$ 56 vol.%	1.3	AZ91	PI	AISI 1018, AISI 4140
[18]	–	2, 5, 10	Pin-on-disc	AlN$_p$ 10, 20 vol.%	–	Mg alloy	SC	Steel, Al$_2$O$_3$
[19]	0.62, 0.94, 1.25	20, 40, 60, 80, 100	Pin-on-disc	Feldspar 1, 3, 5 wt.%	30–50	AZ91	LM	EN24 steel
[20]	0.5	5, 10, 50	Pin-on-disc	SiC$_p$ 30 vol.%	40	Mg	SC	–
[21]	1, 1.5, 2	10, 15, 20	Pin-on-disc	TiC 10 wt.%; MoS$_2$ 10 wt.%	TiC 23; MoS$_2$ 45	Al-7075	SC	EN 32 steel
[22]	0.5, 1, 1.5, 2, 2.5, 3	5, 10, 15, 20, 25, 30	Pin-on-disc	SiC$_p$ 5, 10 wt.%; Gr 5, 10 wt.%	50	Mg	PM	EN31 steel
[23]	0.419	9.8–88.2	Block-on-ring	TiC$_p$ 5, 10, 15 wt.%	5	AZ91	SC	GCr15 bearing steel

SC stir casting, PM powder metallurgy, PI pressureless infiltration, DMD disintegrated melt deposition, LM liquid metallurgy
* mean that units are different to the units included in the title of the column.

AISI 4140, AISI 1018, and H13 steels. Wear resistance was evaluated under dry sliding condition at different loads. In this study it was found that the higher normal load applied, the higher weight loss. In addition, magnesium alloy shows better wear resistance than composite material, especially when normal applied load was 1 GPa; however at 0.5 GPa, differences in weight loss were not so big, which can be seen in the weight loss curves against sliding distances showed in Fig. 7.8. It is presumed that higher loads cause fractures and debonding of high hardness reinforcing particles, increasing wear and starting three-body abrasion that in these particular case contributes to increase in wear. In case of the coefficient of friction, a reduction was observed with decrease in applied load, for both alloy and composites. Decrease in friction is attributed to formation of a lubricating layer of oxides which prevents severe surface wear by providing good morphological stability.

Chemical analyses have shown the creation during the test of different oxides corresponding to the elements present in the composite. Generalized wear mechanisms of the composites are basically type abrasion-adhesion. The wear resistance in all cases was better in the Mg-AZ91E alloy than in the composite Mg-AZ91E/TiC.

Arreola [18] manufactured composites using magnesium alloy AZ91E reinforced with aluminum nitride particles, by stir casting process adding 10, 15, and 20 vol.% AlN_p. The manufactured composites were tested using a pin-on-disc device, under loads of 2, 5, and 10N, against alumina and steel counterparts. In Fig. 7.9 it can be seen that COF decreases with the rise of normal load in both the AZ91E alloy used as matrix and AZ91E/AlN composites, against steel (Fig. 7.9a) and alumina (Fig. 7.9b) counterpart; this may be attributed to the smoothness of surfaces due to the applied load and the increase in tribochemical products.

Due to normal applied load, it is clearly seen in Fig. 7.10 that under higher applied load materials exhibit higher wear, as well as the beneficial effect of reinforcing material at loads of 2 and 5N, since the MMC with greater amount of reinforcement presents a lower weight loss with respect to the matrix and composites with less amount of reinforcement when these were tested against both counterparts.

However, at greater amount of reinforcement and under a load of 10N, the beneficial effect of reinforcement is reversed, since material loss is almost equal. When the counterpart is the steel ball or greater than the magnesium alloy, when tested against the alumina sphere, these results agree with those reported by Alpas and Zhang [24].

In this work, it is established that in all samples, abrasion mechanism is present, being dominant in composites with low amounts of reinforcement and in the Mg alloy. Figure 7.11a shows a wear print. By increasing amounts of reinforcement into the matrix, tribochemical products generation increased. Thus it is presumed that wear mechanism in these materials was tribochemical reaction, followed by abrasion, which can be observed in Fig. 7.11b.

Analysis of wear in a magnesium composite with SiC reinforcement manufactured by mechanical disintegration deposition technique [10] found that 26.3 wt.% provides good wear resistance in comparison with the Mg matrix; however, the highest specific wear rate was reached under the lowest load. The explanation is that under normal loads over 2N tribochemical products were formed

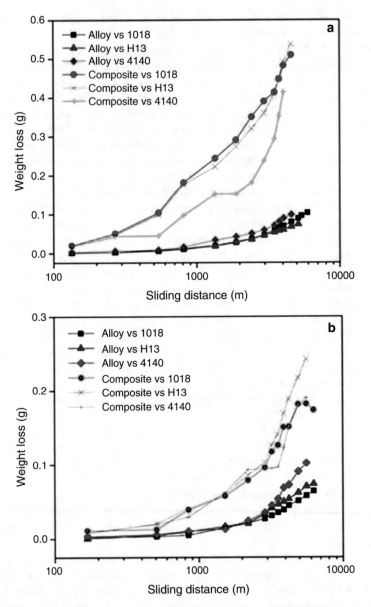

Fig. 7.8 Weight losses of worn samples in AZ91E/TiC composite and alloy AZ91E, (**a**) at 1.0 MPa and (**b**) at 0.5 MPa [17]

as form of dense layers of magnesium hydrous silicate, which softens the contact between the parts, thus reducing wear when applied loads are higher than 2N, as far as coefficients of friction did not observe significant differences between MMC and the alloy. On the other hand, it was found that under normal load of 10N, the COF

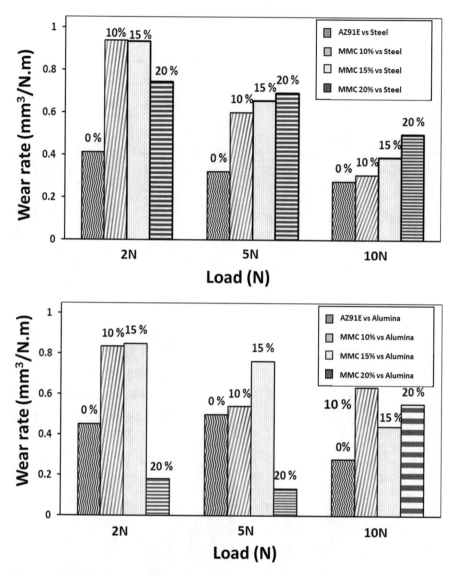

Fig. 7.9 COF in AZ91E alloy, and AZ91E/AlN composite vs (**a**) steel and (**b**) alumina sphere [18]

decreases with respect to 2N load in composites with 26.3 wt.% reinforcement. Also in studies carried out by Sharma et al. [19], it was found that wear rate in composites made with magnesium alloy matrix and feldspar reinforcements decreases at higher reinforcement content and increases at higher applied load, in both composites and matrix alloy. They suggest that at lower loads the fractured feldspar forms a layer at interface that protects the matrix, avoiding metal-composite contact, but at higher loads the formed layer is fractured allowing counterpart to come into direct contact

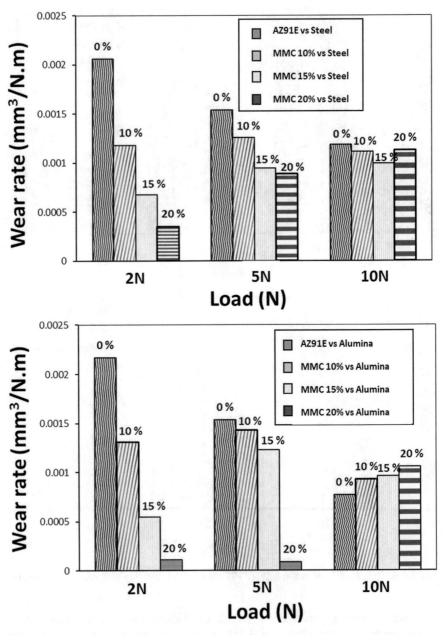

Fig. 7.10 Wear rate in AZ91E alloy and AZ91E/AlN composite, vs (**a**) steel and (**b**) alumina sphere [18]

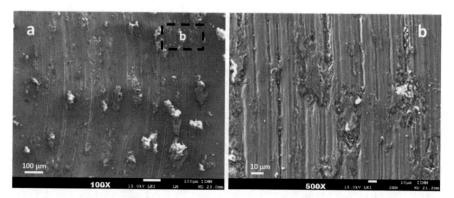

Fig. 7.11 Wear track in MMC AZ91E/AlN 20 vol.%: (**a**) 100× and (**b**) 500× [18]

with the matrix where high strains are developed, promoting the removal of superficial layers by delamination. Saravanan et al. [20] made a pure magnesium matrix composite with 30 vol.% SiC reinforcements.

The conclusions in this work were that SiC reinforcement improves wear resistance up to two orders of magnitude compared to matrix; these results contrast with those found by Ramakoteswara et al. [12], who suggest that amounts greater than 8 wt.% of reinforcement no longer improve wear resistance. Likewise, the conclusions obtained by Alahelisten et al. [25], who manufactured composites with magnesium and aluminum matrices up to 30 wt.% of alumina fibers by squeeze-casting process, found that 10% fiber reinforcement of alumina is the optimum amount. As for the applied load, researchers found that at higher applied load the composites undergo greater wear. In other investigations, it has been found that under certain conditions the increase of normal applied load decreases wear rate of tested materials.

Asif et al. [15] found that in a composite made of aluminum and SiC reinforcements with 20% wt., wear rate increases with the increase in applied normal load, but in composites made with aluminum reinforced with SiC, graphite, and Sb_2S_3, the behavior was different. Since applying loads from 30 to 50N, wear rate increases under sliding speeds of 5 and 7 m/s, respectively, but at normal loads between 50 and 80N, the wear rate decreases even at sliding speeds of 7 m/s. Wear rate is lower under loads of 80N compared to loads of 30N. At 5 m/s of sliding speed, a slight decrease in the wear rate was observed with all applied loads, attributing this effect to the work surface hardening, oxide formation, and crushing of silicon carbide particles as well as graphite smearing on surfaces. They report that the coefficient of friction usually decreases with load, both in composite reinforced with SiC and in that manufactured with SiC plus graphite lubricant.

Narayanasamy et al. [21] studied behavior in composites made with an aluminum alloy reinforced with titanium carbide and MoS_2 as solid lubricant; to design the experiments, they used the Taguchi method with an orthogonal arrangement L27. The conclusions of these works were that Taguchi technique can be used to predict wear rate of composites made with aluminum alloy 7075 reinforced with TiC and

MoS_2 by stir casting process and that wear rate of composites increases with the applied load, sliding speed, and sliding distance. Researchers also found that composites with the addition of MoS_2 present better wear resistance than those without reinforcement. Prakash et al. [22] investigated wear resistance of composites made of magnesium reinforced with SiC particles (5 and 10 wt.%) and as lubricant graphite in several quantities (5 and 10 wt.%). Results indicate that composites with higher amount of reinforcement and 5% of graphite show the best wear resistance with different normal applied loads and sliding speeds, although they found that the higher normal load, the higher wear and the higher sliding speed, the minor weight loss. The COF increases with the SiC; nevertheless, the addition of graphite reduces it; on the other hand, they report that the higher load the higher COF, but at higher sliding speed, a reduction in the coefficient of friction was observed; it was attributed to the formation of a mechanical mixed layer on the counterpart disc. They point out that even though the composites with the highest amount of SiC and 5% of graphite has a lower hardness than the composites with 10% SiC, it shows a better wear behavior. Xiu et al. [23] studied the behavior of composites made with AZ91 magnesium alloy reinforced with TiC particles fabricated by melting and stirring. Results showed that composites with greater amount of reinforcement showed better wear resistance than those with less reinforcement and even more than the matrix and likewise found that at higher loads composites and matrix exhibit greater material lost. On the other hand, and differing with other researches, it was reported that with higher applied loads the coefficient of friction increases due to the increase of surface roughness and also found higher coefficient of friction in the alloy compared to the composites.

7.6.2   Influence of Sliding Speed on MMC Wear

A large number of studies have been carried out about sliding speed influence over wear, while keeping the applied load constant. Sliding speed causes a temperature increase with time, which allows to occur chemical reactions or interdiffusion between the bodies in contact, forming layers of reaction products on the interface known as mechanical mixed layer (MML). The properties of these layers depend on both, the coefficient of friction and wear rate, since these layers can be hard or soft and depend on the elements that conform the sample and counterpart; if the formed layer is fragile, when the load increases, it can fracture and allow mechanical contact between the two bodies of the tribosystem, which would increase friction and wear. Also, it is very important to take into consideration environment in tribological pair, that is, humidity, temperature, and gases; likewise high temperatures generated by sliding and normal applied load can generate phase transformations and modify mechanical properties. Table 7.2 shows manufacturing processes, matrices, and reinforcements, as well as conditions under which some researchers performed different tests. Basavarajappa et al. [26] studied the influence of sliding speed on wear in composites made of aluminum alloy and reinforced with SiC particles and

Table 7.2 Research about sliding speed influence in wear

Ref.	Sliding speed (ms⁻¹)	Normal load (N)	Apparatus	Reinf. type	Reinf. size (μm)	Matrix type	Fabrication	Counterpart type
[14]	1, 3, 5, 7, 10	10, 30	Pin-on-disc	Al_2O_{3p} 0.66, 1.11, 1.50 vol.%	50 nm*	AZ31B	DMD	Oil-hardened tool steel
[16]	0.6, 0.9, 1.2	5, 7.5, 10	Pin-on-disc	ZnO 0.5 vol.%	50–200 nm*	Mg	PM	EN36 steel
[26]	1.53, 3, 4.6, 6.1	40	Pin-on-disc	SiC_p 15 wt.%; SiC_p 15, Gr 3 wt.%	25	AA2219	LM	EN36 steel
[27]	1.2, 2.5, 3.7, 5.1	10, 20, 30, 40	Pin-on-disc	SiC_p 1, 2, 3, 4, 5 wt.%	30	ZA43	LM	Hard steel disc
[28]	1, 3, 5, 7, 10	10	Pin-on-disc	Al_2O_{3p} 1.5 vol.%	50 nm*	AZ31B	DMD	AISI-O1 tool steel disc
[29]	1, 3, 5, 7, 10	10	Pin-on-disc	Al_2O_{3p} 0.22, 0.66, 1.11 vol.%	50 nm*	Mg	DMD	AISI-O1 tool steel disc
[30]	1.5, 3, 4.5	9.8, 29.4, 49.1	Pin-on-disc	TiC_p 3, 4, 5, 6, 7%	–	AA6061	SC	EN32 steel

SC stir casting, *PM* powder metallurgy, *LM* liquid metallurgy, *DMD* disintegrated melt deposition
* mean that units are different to the units included in the title of the column.

3 wt.% graphite as lubricant. They found that addition of SiC in the alloy increases wear resistance of composites at 4.6 m/s sliding speed, which is attributed to the formation of a layer consisting in aluminum oxides and iron oxides from counterpart and silicon carbide particles, but at higher speeds than 4.6 m/s wear rate increases slightly in composites and abruptly in matrix alloy. Research performed by Selvam et al. [16] found that composite material undergoes higher wear with increase in speed at all normal loads tested.

It was also found that changes in the wear rate at the speed between 0.6 m/s and 0.9 m/s increased, but at the speed of 1.2 m/s the rate presents a decrement, which is attributed to oxide formation.

Rajaneesh et al. [27] conducted a research about wear behavior in composites made of an aluminum-zinc alloy reinforced with SiC particles. It was conclude that wear rate decreases with the increase of reinforcing particles and rises with the increase of sliding speed as well as with the increase in normal load. They also observed that due to friction, there was an increase in temperature at interface causing softening and delamination; another mechanism reported is abrasion. Sharma et al. [19] conclude that by increasing sliding speed in magnesium-reinforced composites and feldspar particulate reinforcement, wear rate increases because of formed tribo-layer cracks at high speeds. Likewise, they state that mail wear registered in the alloy at low speeds is attributed to oxidative mechanism. Nguyen et al. [14] investigated the influence of sliding speed on wear rates and friction of composites made by magnesium alloy and alumina nanoparticle reinforcement and reported that at low sliding speeds the matrix alloy shows better wear resistance than nanocomposites, but, at higher test speeds, the behavior change and the alloy exhibit greater wear.

On the other hand, they found that at normal applied load of 10N, the lowest wear rate was presented at 5 m/s and under 30N the lowest wear rate was reached at a speed of 3 m/s; the lower coefficient of friction for both the alloy and composite was also achieved at 3 m/s under 30N, respectively. The wear mechanisms that occurred in all specimens were abrasion, oxidation, adhesion, softening, and melting, but in the case of composites, delamination was also present. Shanthi et al. [28] studied composites made with a magnesium alloy reinforced with Al_2O_3 nanoparticles with additions of calcium; they found that the best wear resistance occurred due to the increase in hardness and strength because of the addition of calcium and the formation of hard intermetallic (Mg, Al) 2Ca. They also found that wear rate in composites, especially when they added 3 vol.% Ca, decreases consistently with sliding speed. As for wear mechanisms, they found that at low speeds and 10N load, abrasion predominates, in agreement with Arreola [18] and Sharma et al. [19] who observed that the predominant mechanism at low sliding speeds and loads of 10N was abrasion; thermal softening mechanism was only present in the alloy. Lim et al. [29] performed a study about wear in a pure Mg matrix composite, reinforced with alumina nanoparticles, and also observed that at low loads and speeds smaller than 7 m/s the main wear mechanism was abrasion and at higher speeds the temperature and plastic deformation increase, resulting in a change in wear mechanism from cutting to plowing or wedge formation, which is in agreement with Shanthi et al. [28].

7.6.3   Influence of Size, Shape, and Amount of Reinforcement

Although there is controversy over the amount of reinforcement to obtain the greatest wear resistance of different matrices, some researchers propose that at high amounts of reinforcement, wear resistance decreases because with greater amount of reinforcement coexists higher number of interfacial area between reinforcement and matrix where exist stress concentration and cracks are generated and propagated. Table 7.3 shows manufacturing processes, matrices, and reinforcements, as well as conditions under which some researchers performed different wear tests. In a research made by Lakshmipathy et al. [31], about wear on aluminum alloy composites reinforced with SiC and Al_2O_3, they found that volume lost was smaller with the increase in reinforcement but increased by increasing normal load. They report that the coefficient of friction in alloys was smaller than in composites, which is in agreement with Ramírez [32].

They also explain that the COF decreases with the number of strokes since at greater sliding distances the temperature over surfaces increases, resulting in surface softening. In contrast to previous results, Gopalakrishnan et al. [30] found that increasing reinforcement in aluminum alloy composites increases specific strength but reduces wear resistance, which seems surprising since the reinforcement volumes were between 3 and 7% and neither found an explanation for this behavior in MMC wear. Miyajima et al. [33] studied wear behavior of composites made with aluminum alloys and reinforced with different forms (particles, fibers, and whiskers) of alumina and silicon carbide. It was found that wear behavior of composites depends into a large extent on the shape of reinforcement and volume added into the matrix; also they revealed the optimum amount of reinforcements, 22, 10, and 2 vol.%, for whiskers, fibers, and particles, respectively, although in the case of particles it was observed that between 2 and 10 vol.% of reinforcement there was a slight decrease in wear rate. They conclude that particle reinforcements provide better benefits than fibers and whiskers. Zou et al. [34] report a significant decrease in wear of composite materials made from an aluminum alloy and reinforcements up to SiC 50 vol.% compared to matrix and minor reinforcement additions. Similarly, they report that wear resistance is proportional to both reinforcement volume and particle size; MMCs with highest wear resistance were composites with 38% reinforcement and 57 μm, although results are not reported with 50 vol.% and 57 μm; it can be seen that the best wear resistance was obtained with 50 vol.% reinforcement. They propose as wear mechanisms plastic deformation, particle fracture, and delamination. These results contrast with those reported by Ramakoteswara [12] who found an optimum amount of reinforcement; on the other hand, the composites Al-2024/TiC_p with 52 vol.% of reinforcement registered a better wear resistance, with and without heat treatment [32]. As for particle size, these results agree with those reported by Maleque et al. [35] who found that reinforcement additions of 80 μm average into aluminum alloy exhibit better wear resistance than adding 40 and 15 μm particles. Kok and Ozdin [9] carried out a study on wear of MMC formed by an aluminum alloy and different amounts of Al_2O_3

Table 7.3 Research about amount of reinforcement influence in wear

Ref.	Sliding speed (ms^{-1})	Normal load (N)	Apparatus	Reinf. type	Reinf. size (µm)	Matrix type	Fabrication	Counterpart type
[9]	2	2, 5	Pin-on-disc	Al_2O_{3p} 10, 20, 30 wt.%	16, 32	Al-2024	V	SiC emery paper
[12]	2	10, 20, 30	Pin-on-disc	TiC_p 2, 4, 6, 8, 10 wt.%	2	AA7075	SC	EN 32 steel
[20]	0.5	5, 10, 50	Pin-on-disc	SiC_p 30 vol.%	40	Mg	SC	–
[25]	0.83	50	Block-on-ring	δ-Al_2O_{3f} 10, 20, 30 vol.%	Diameter 3; length 500	Mg, Al, Mg-Al alloy	SqC	Carbon steel
[31]	–	20, 50, 75	Reciprocating machine	$Al7075$-SiC_p $Al6061$-Al_2O_{3p} (10, 15, 20% both SiC and Al_2O_3)	36 (both SiC and Al_2O_3)	Al7075 Al6061	SC	EN32 steel
[33]	0.1	10	Pin-on-disc	SiC_w: 5–29 vol.% SiC_p: 2–10 vol.% Al_2O_{3f}: 3–26 vol.%	SiC_w: $d = 0.3$–1 $l = 5$–15 SiC_p: 10 Al_2O_{3f}: $d = 4$, $l = 40$–200	Al 2024, Al ADC12	MMC_p-PM MMC_w-HPI MMC_f-HPI	0.45% carbon steel
[34]	0.94	55	Pin-on-ring	SiC_p 5, 13, 38, 50 vol.%	5.5, 11.5, 57	Al-Si-Cu alloy	IP and SC	SAE 52100 steel
[35]	–	–	Pin-on-disc	SiC_p 9 wt.%	15, 40, 80	Al 6061	LM	–

SC stir casting, *SqC* squeeze casting, *V* vortex method, *PM* powder metallurgy, *HPI* high pressure infiltration, *LM* liquid metallurgy, *IP* infiltration process

reinforcement under loads of 2 and 5N, finding that lost volume in MMC was significantly reduced by increasing the amount of reinforcement up to 30 wt.% and particle size. Similarly, they indicate that the volume loss in studied materials increases with normal applied load, distance, and particle size of the SiC abrasive paper used in these experiments. They report that the wear mechanisms presented in this research were microcutting and microplowing.

7.6.4 Effects of Heat Treatments in MMC on Wear Resistance

The effects of heat treatments on wear resistance in metallic matrix composites have been studied [32, 36], finding that one of main factors influencing wear resistance is the formation of hard intermetallic phases that increase composite hardness. Chelliah et al. [36] investigated the effect of heat treatments on composites made with a magnesium alloy and TiC reinforcements, resulting that heat-treated (HT) MMC undergoes a higher wear than the non-heat-treated composites, arguing that the difference between thermally treated material and as cast composite is due to the fact that in the case of HT composite, the β-phase decreases; therefore composites are more susceptible to oxidation than as cast, and magnesium oxide formed is fragile, which fractures with high applied loads, and metal/metal contact happens, causing greater wear. They found that the COF in HT MMC decreases up to 4.5 times related to the as cast, which explains this as a function of oxide layer formed in HT composite. They do not explain the reduction in hardness of HT composite. This result contrasts with those obtained by [37–40] who applied a HT to MMC obtaining an increase in hardness and wear resistance. Kaczmar and Naplocha [37] studied wear resistance in composites made of an aluminum alloy and reinforcements with δ alumina fibers; composites were subjected to T6 treatment; results reveal that T6 HT composites register a reduction in wear rate up to three times than the alloy at normal load of 0.8 MPa, nevertheless increasing the load to 1.2 MPa; the HT alloy registers greater wear resistance; it is attributed to rupture and exfoliation of fibers.

Ramírez [32] investigated behavior of composites made with 2024 aluminum alloy reinforced with 53 wt.% TiC particles by infiltration process without external pressure and subjected to a T6 heat treatment. Results are presented in Fig. 7.12 where is clear that HT composites showed higher wear resistance than those as-fabricated and that the matrix, both using steel counterpart (Fig. 7.12a); as well as those tested against alumina ball (Fig. 7.12b). On the other hand, the matrix exhibits lower COF than the hardest composites. It can also be seen in Fig. 7.12 that with the load increase wear rates increases, being this more noticeable in the matrix.

Figure 7.13 shows micrographs of worn surfaces in matrix and thermally treated composites; it can be seen in Fig. 7.13a that the grooves are wider and deeper than those observed in Fig. 7.12b and can also be appreciated in Fig. 7.13b a layer of tribochemical products identified as aluminum and titanium oxides. An increase in tribochemical products is observed with the increase of load and sliding distance.

Fig. 7.12 Variation of wear
rate for Al-2024 and
Al-2024/TiC composites
against (**a**) steel ball and (**b**)
alumina ball, at different
loads [32]

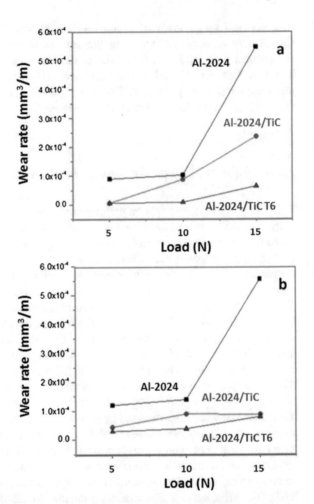

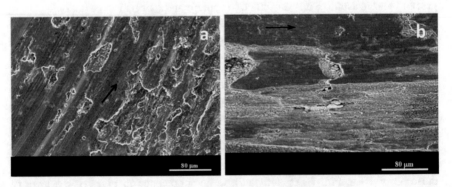

Fig. 7.13 Micrograph of (**a**) worn surface of the alloy showing the grooves and (**b**) worn surface of
Al-2024/TiC HT showing the oxide formation [32]

The increase in the wear resistance of Al-2024/TiC composites is attributed to the formation and diffusion of reaction products on the metal surface, which decreases and softens the contact area during sliding, as well as the hardening presented because of the applied heat treatment.

7.6.5 Influence of Fabrication Methods and Sliding Distance on Wear Resistence

MMC manufacturing process has been found to influence wear resistance of these. Suresh et al. [41] compared weight loss of two Beryl/Al-Si-Mg composites manufactured by squeeze-casting and gravity cast method; they concluded that squeeze-casting technique shows better wear resistance since by this technique it is possible to manufacture MMC with higher density and hardness. Lim et al. [42] compared wear resistance of two Al-Cu/SiC$_p$ composites made by powder metal-lurgy (PM) and rheocast routes, which were also compared against the alloy without reinforcements, concluding that MMC made by rheocast technique presented better wear resistance than those made by PM, and the matrix alloy manufactured by the same rheocast process exhibits greater wear resistance than those manufactured by PM technique at loads greater than 40N. On the other hand, Sahin [43] investigated wear behavior of SiC particles reinforcing aluminum composites using pin-on-disc tests and Al$_2$O$_3$ and SiC emery paper as counterparts with different particle sizes, finding that composites exhibit higher wear resistance than the matrix, when normal load, particle size of the emery paper, or test distance increase, but in all cases also the wear increases, except when Al$_2$O$_3$ abrasive paper is used as counterpart. The decrease in wear rate with the distance is attributed to a cover and clogging of the surface, as well as a hardening by mechanical work due to longer distances com-bined with high normal load and sliding speed.

References

1. Jost HP (1976) Economic impact of tribology. In: Proceeding of the 20th Meeting of the Mechanical Failures Prevention Group
2. Pinkus O, Wilcock DF (1997) Strategy for energy conservation through tribology. In: Tribology in Energy Technology Workshop American Society of Mechanical Engineers
3. Holmberg K, Anderson P, Erdemir A (2012) Global energy consumption due to friction in passenger cars. Tribol Int 47:221–234
4. Rabinowicz E (1995) Friction and wear of materials. Wiley, New York
5. Archard J (1953) Contact and rubbing of flat surfaces. J Appl Phys 24(8):981–988
6. Hutchings IM (1992) Tribology: friction and wear of engineering materials. BH, p 112
7. Glossary of Terms (1992) ASM handbook, friction, lubrication and wear technology. ASM Int 18:21
8. Brushan B (2013) Introduction to tribology, 2nd edn. Wiley, New York

9. Kok M, Ozdin K (2007) Wear resistance of aluminum alloy and its composites reinforced by Al_2O_3. J Mater Process Technol 183:301–309
10. Manoj B, Basu B, Murthy V et al (2005) The role of tribochemistry on fretting wear of Mg-SiC particulate. Compos Part A 36:13–23
11. Suh N (1973) The delamination theory of wear. Wear 25:111
12. Ramakoteswara V, Ramanaiah M, Sarcar M (2016) Dry sliding wear behavior of TiC-AA7075 metal matrix composites. Int J Appl Sci Eng 14(1):27–37
13. Lim C, Lim S, Gupta M (2003) Wear behaviour of SiC_p-reinforced magnesium matrix composites. Wear 255:629–637
14. Nguyen Q, Sim Y, Gupta M, Lim C (2014) Tribology characteristics of magnesium alloy AZ31B and its composites. Tribol Int Part B 82:464–471
15. Asif M, Chandra K, Misra P (2011) Development of aluminum hybrid metal matrix composites for heavy duty applications. J Miner Mater Charact Eng 10(14):1337–1344
16. Selvam B, Marimuthu P, Narayanasamy R et al (2014) Dry sliding wear behavior of zinc oxide reinforced magnesium matrix nano-composites. Mater Des 58:475–481
17. Falcón L, Bedolla E, Lemus J (2011) Wear performance of TiC as reinforcement of a magnesium alloy matrix composite. Compos Part B 42:275–279
18. Arreola C (2016) Evaluación de propiedades mecánicas y comportamiento al desgaste de compuestos AZ91E/AlN fabricados por fundición con agitación. Master Thesis, Instituto Investigación Metalurgia Materiales, UMSNH, México
19. Sharma S, Andand B, Krishna M (2000) Evaluation of sliding wear behavior of feldspar particle-reinforced magnesium alloy composites. Wear 241:33–40
20. Saravanan R, Surappa M (2000) Fabrication and characterization of pure magnesium-30 vol. % SiC_p particle composite. Mater Sci Eng A276:108–116
21. Narayanasamy P, Selvakumar N, Balasundar P (2015) Effect of hybridizing MoS_2 on the tribological behaviour of Mg–TiC composites. Trans Indian Inst Met 68:911–925
22. Prakash K, Balasundar P, Nagaraja S et al (2016) Mechanical and wear behaviour of Mg-SiC-Gr hybrid composites. J Magnes Alloys 4:197–206
23. Xiu K, Wang HY, Sui HL et al (2006) The sliding wear behavior of TiC/AZ91 magnesium matrix composites. J Mater 41:7052–7058
24. Alpas A, Zhang J (1992) Effect of SiC particulate reinforcement on the dry sliding wear of aluminium-silicon alloys (A356). Wear 155:83–104
25. Alahelisten A, Bergman F, Olsson M, Hogmark S (1993) On the wear of aluminium and magnesium metal matrix composites. Wear 165:221–226
26. Basavarajappa S, Chandramohan G, Mahadevan A (2007) Influence of speed on the dry sliding wear behavior and subsurface deformation on hybrid metal matrix composite. Wear 262:1007–1012
27. Rajaneesh N, Sadashivappa K (2011) Dry sliding wear behavior of SiC particles reinforced zinc-aluminium (ZA43) alloy metal matrix composites. J Miner Mater Charact Eng 10 (5):419–425
28. Shanthi M, Lim C, Lu L (2007) Effects of grain size on the wear of recycled AZ91 Mg. Tribol Int 40:335–338
29. Lim C, Leo D, Gupta M (2005) Wear of magnesium composites reinforced with nano-sized alumina particulates. Wear 259:620–625
30. Gopalakrishnan S, Murugan N (2012) Production and wear characterization of AA 6061 matrix titanium carbide particle reinforced composite by enhanced stir casting method. Compos B 43:302–308
31. Lakshmipathy J, Kulendran B (2014) Reciprocating wear behaviour of 7075Al/SiC and 6061/ Al_2O_3 composites: a study of effect of reinforcement, stroke and load. Tribol Ind 36(2):117–126
32. Ramírez REJ (2015) Thesis: Efecto del tratamiento térmico T6 sobre las propiedades tribológicas del compuesto Al-2024/TiC, Tesis Universidad Autónoma de Coahuila
33. Miyajima T, Iwai Y (2003) Effects of reinforcements on sliding wear behaviour of aluminium matrix composites. Wear 255:606–616

34. Zou X, Miyahara H, Yamamoto K et al (2003) Sliding wear behaviour of Al-Si-Cu composites reinforced with SiC particles. Mater Sci Technol 19(11):1519–1526
35. Maleque M, Radhi M, Rahman M (2016) Wear study of Mg-SiC$_p$ reinforcement aluminium metal matrix composite. J Mech Eng Sci 10:1758–1764
36. Chelliah N, Singh H, Surappa M (2016) Correlation between microstructure and wear behavior of AZX915 Mg-alloy reinforced with 12 wt% TiC particles by stir-casting process. J Magnes Alloy 4:306–313
37. Kaczmar J, Naplocha K (2010) Wear behavior of composite materials based on 2024 Al-alloy reinforced with δ-alumina fibers. J Achiev Mater Manuf Eng 43:8–93
38. Shivaprakash Y, Basavaraj Y, Sreenivasa K (2013) Comparative study of tribological characteristics of AA2024+10% fly ash composite in non-heat treated and heat treated conditions. Int J Res Eng Technol 2:175–280
39. Sameezadeh M, Emamy M, Farhangi H (2011) Effects of particulate reinforcement and heat treatment on the hardness and wear properties of AA 2024-MoSi$_2$ nanocomposites. Mater Des 32:2157–2164
40. Yamanoglu R, Karakulak E, Zeren A et al (2013) Effect of heat treatment on the tribological properties of Al-Cu-Mg/nano SiC composites. Mater Des 49:820–825
41. Suresh K, Niranjan B, Jebaraj M et al (2003) Tensile and wear properties of aluminium composites. Wear 255:638–642
42. Lim SC, Gupta M, Ren L (1999) The tribological properties of Al-Cu/SiC$_p$ metal matrix composites fabricated using the rheocasting technique. J Mater Process Technol 89–90:591–596
43. Sahin Y (2003) Wear behavior of aluminum alloy and its composites reinforced by SiC particles using statistical analysis. Mater Des 24:95–103

Index